Climate Change Adaptation in 2011

A Selection of Writings, Blog Posts and Links
by Brian Thomas

Carbon Based Press

Climate Change Adaptation in 2011:
A Selection of Writings and Links

Published by the Carbon Based Press
PO Box 234
West Cornwall, Connecticut 06796

Dedicated to
Catherine Noren

Introduction

Climate change adaptation teems with large events and subtle forces. Every day, the news brings stories of climate-related floods, droughts, forest fires, sea level rise, extreme weather, disasters, ecological stress, and infectious diseases. The field also spans new developments in computer modeling, agriculture, water and land use, paleoclimate, scientific research, insurance, and finance. Climate instability takes a dramatic toll on indigenous peoples and developing countries, but it also warps the psychological and social quality of our lives, and in my blogging I don't neglect these subjects, either. You'll find Carbon Based at http://carbon-based-ghg.blogspot.com.

I link to five news stories daily, and I still miss plenty, so I make no claim to completeness. Daily blogging heightens my appreciation for the scale and variety of climate-related impacts.

The scale and variety are so large, in fact, that I narrow my focus. I avoid many stories that are well-covered by other bloggers, such as those dealing with cutting greenhouse gas emissions, or electoral politics and climate change.

Excerpting my writings, blog posts and link for this book format restricts the available space even more. To make this publication a manageable size, I've eliminated more categories. No stories about nitrogen pollution, even though it's a mainstay of the blog, and no pictures of polar regions, even though I love to post them.

For this year's annual review, I've focused on stories about climate risks, since I view risk as the one way of unifying the thronging variety of climate change adaptation. It's a central topic because one effect of global warming is worsening probabilities, something that is notoriously difficult to detect. It's a natural filter for me because risk has been the subject of many of the articles I've written during 2011.

I begin this book with my own writings, both in print and on my blog, and then include a sample of links about risk topics, often with an insurance angle.

"Insurance, cognitive bias, and the struggle against climate risks" appeared in the Fall 2011 edition of *The John Liner Review*, a quarterly journal of "advanced risk management strategies." It summarizes and brings together several recent themes of my work, focusing on the psychological and social obstacles to taking climate change seriously. I believe that cognitive biases help explain the dangerous persistence of climate change denial, and why overcoming it will be so difficult. I also explore some instances of these forces at work in the insurance industry, discussing Florida's disaster insurance program, Citizens Property, and the challenges facing flood insurance in the United States, where we spend a lot of money encouraging people to build valuable property in vulnerable flood plains. Relentless political propaganda only accounts for part of the trouble.

"Climate and bondage" builds on some recent philosophy books that explore the moral implications of climate change, such as Stephen Gardiner's *A Perfect Moral Storm*. Here I make a hopeful analogy to the struggle against runaway greenhouse gas emissions and the decades-long campaign to abolish slavery. It's an open question whether a 21st century Harriet Beecher Stowe is out there, writing the book that will galvanize public opinion. I believe that one day future generations will read the claims of climate change denialists with the same disgusted astonishment that today's people read the pro-slavery writings that preceded the Civil War in the United States. Climate change poses a moral risk because our accustomed practices raise the likelihood of grave consequences.

"What lasts?" is a business-oriented piece about sustainability written for a client but never used and published for the first time here. It weaves together various kinds of sustainability, and finds the label more complex than is commonly believed. Here I raise a question that also looms large in my blog -- how we are going to deal with restoring our long-neglected infrastructure in the United States (and elsewhere)?

Hurricane Irene and other storms prompted many of my personal blog posts in 2011. I wrote about a neighbor, Tim Prentice, whose kinetic sculptures are moved by the wind, investigating how his airy pieces hold up in a gale. It's a non-esthetic approach to an artistic subject, but Tim's sculptures lend themselves to it because they are as fragile-looking as they are beautiful. Like humans in general, they are tougher than they appear.

The failed dam in my yard provided a few paragraphs of fretting, in which my inability to fix the damned thing serves as a microcosm for society's response to climate change.

My experiences with our week-long blackout in Connecticut before and after Halloween was a fruitful source of topics. I must admit that my personal adaptation record is mixed. You don't appreciate how much you depend on electricity until it's gone.

I'm also pleased with the post I did about John Ruskin's 1884 lecture, "The Storm-Cloud of the 19th Century," a nutty but prophetic examination of the effect that humans have upon the weather and the climate. Ruskin comes to no conclusions, really, and is hampered by his attempts to deal with a scientific topic in an poetic way, but the result is surprisingly pertinent. I wonder how he would have updated *The Stones of Venice?*

The links I've selected here continue the risk theme. You'll find many stories about natural catastrophes and their costs to society and business, as well as land use decisions that carry unexpected impacts. One of the most significant came early in the year, when the town of Ventura, California decided to let some beach infrastructure degrade without being replaced, moving the new basketball courts and paths further inland. Perhaps this kind of managed retreat in the face of sea level rise will serve as a model for other communities.

Another wail from Cassandra came in September from the National Institutes of Standards and Technology, which warned of the dangers of considering natural catastrophes in isolation. Trouble often comes in clusters, so that risks appear manageable in theory, but turn out to be anything but in practice. Risks can accumulate without our being aware of the dangers.

In Carbon Based, I spotlight these stealthy risks as well as potential sources of resilience against them. That's my contribution.

Brian Thomas
Carbon Based

Table of Contents

Insurance, cognitive bias, and the struggle against climate risks

Insurers inhabit the crossroads where business and society handle all risks, including climate change. The task is demanding: uncertainty veils the future, and insurers and the general public are fated to rely on judgments about probabilities. Insurers translate probabilities into commercial terms. Then they retail the science to their clients, with varying profitability. This is true whether the insurer is a private company or the state itself.

Most people judge risk badly, which makes dealing with climate change uniquely difficult. We worry too much about minor hazards and are nonchalant about more serious ones. Travelers are often more comfortable on the ground than in the air, even though drivers are more likely to sustain an accident than flyers.

Risk researchers such as Daniel Kahneman and Amos Tversky have found that people are especially inconsistent at interpreting dangers that are slow, long-term and ambiguous. Long term planning is hard. In most of our endeavors, the present far outweighs the future. We may be vaguely aware of trouble ahead, but we ignore it until we are menaced in an urgent, dramatic way. Our short-term bias accentuates this tendency. Markets heighten our short term bias, especially the financial markets. We want our payoffs quickly and have little appetite for living with uncertainty.

Pioneering psychologists and behavioral economists have explored the cognitive biases that cloud our risk judgment. Some of the most pertinent in climate discussions are:

- Confirmation bias—searching only for data that confirms one's preconceived hypothesis. People do not want global warming to happen and so they seize on apparent disconfirming instances.
- Hyperbolic discounting —Preferring short-term payoff to longer-term larger payoffs, a particularly important tendency. In climate discussions, this results in minimizing the claims of future generations to the world's resources. This largely results from importing a common financial technique —discounting — into the moral realm, where it does not belong.
- Neglect of probability—the tendency to ignore probability when making a decision in the face of uncertainty. This neglect is illustrated every time a developer builds on land that is vulnerable to storm surges or flooding.
- Zero-risk bias—reducing a small risk to zero rather than going for a larger reduction in larger risk. People will spend far too much time trying to reduce a 1 percent risk to zero, rather than the much more valuable step of, say, reducing a risk of 40 percent to 10 percent.

- Focusing effect — putting too much emphasis on a single aspect of a situation or event. In climate discussions, people will pay attention to the few locations where the temperature has fallen and ignore the rest of the world.
- Framing — viewing through a perspective or approach that is too narrow. This usually takes the form of a rigidly economic point of view, ignoring the role that natural resources play as the ultimate foundation of the world's wealth.
- The bandwagon effect—everybody's doing it, or, in the case of climate change, doubting it.

These pervasive biases are well documented. They pose a major obstacle to deciding well about climate change, critical infrastructure failure, and economic volatility[1]. Their cumulative effect steers most individuals and groups making decisions towards "spasmodic, risk-averse, piecemeal approaches."

If we fail to examine and challenge these biases, we are in danger of considering only a truncated range of timid options. In the words of one analyst, "The result can be poor decision process, poor decision quality, ineffective action, blindness to change, and a reversion to the status quo in the face of even the most grave dangers."

The insurance industry exists because of this erratic grasp on risk. For many hazards, insurers understand and assess the risks better than the general public. In theory, they study the actual probabilities of known hazards, figure out a viable premium that gives themselves a profit and the policyholders the agreed upon protection against the risk.

At a certain level, the general public understands the importance of relying on science rather than our gut feelings. We endure the slow, small pain of insurance premiums, for the reassurance of the large compensation should a hazard actually occur. Buying insurance goes against our intuitive feel for risks, but we receive a little more certainty and reduced volatility with our premiums. The insurers make money because they thrive on our shaky sense of the odds.

Climate change risk, one of the chronic, lumbering dangers that activates our cognitive biases, poses special challenges to insurers, not merely because they are exposed to losses from many weather risks such as hurricanes. Insurers are people, too, and the cognitive blind spots that afflict individuals also affect the risk business.

Insurers have some historically enshrined tools for assessing all risks, whether as large as climate change, or as small as house fires. One is simple iteration: insurers check their portfolios regularly for risks that have increased or diminished. They know that risk is a moving target, and accuracy demands continuous attention. If a high-payout event like Hurricane Andrew in 1992 shows them that they have underestimated the dangers of building on vulnerable coastlines, they promptly revise their underwriting. This institutional awareness of the industry's own fallibility is one of the historical strengths of the insurance business.

---

[1] Noah Raford, "Overcoming cognitive bias in decision making," September 14, 2009, http://www.noahraford.com/files/raford_overcomingbias.pdf

Climate Change Adaptation in 2011

Climate risk

Risk lies squarely at the center of climate change, according to the Intergovernmental Panel on Climate Change, the group of scientists assembled by the United Nations to evaluable the risks greenhouse gases pose to the planet. Responding to climate change involves an iterative risk management process that includes both adaptation and mitigation, and takes into account climate change damages, co-benefits, sustainability, equity and attitudes to risk.[2]

The dangers are enormous: The IPCC Fourth Assessment Report (2007) as well as several other more recent studies summarize the broad evidence for global warming. An updated summary of the impacts comes from the Critical Decade, published by the Australian government's Climate Commission[3].

- The average air temperature at the Earth's surface continues to rise at a rate of $0.17^\circ C$ per decade over the past three decades.

- The upper 700 meters of the ocean are storing most of the planet's excess heat, and water temperatures continue to increase.

- Anthropogenic CO2 emissions are acidifying the world's oceans.

- Recent observations confirm net loss of ice from the Greenland and West Antarctic ice sheets; the extent of Arctic sea ice cover continues on a long-term downward trend. Most land-based glaciers and ice caps are in retreat.

- Sea-level has risen at a higher rate over the past two decades, consistent with ocean warming and an increasing contribution from the large polar ice sheets. The rise is at the very high end of IPCC estimates.

- Arctic sea ice is declining far faster than IPCC models have projected, currently approximately 40 years ahead of schedule.

The world's biological systems are behaving exactly as predicted by the theory of a warming Earth, with observed changes in gene pools, the ranges of plants and animals, the sequence of biological patterns and ecosystem dynamics. The Australian report (like every other major scientific body that has studied the issue) notes that the past decade (2001-2010) was the hottest on record, $0.46^\circ C$ above the 1961-1990 average.

---

[2] The Intergovernmental Panel on Climate Change Fourth Assessment Report (2007), page 22.
[3] The Australian Government's Climate Commission, "The Critical Decade: Climate Science, Risks and Responses," http://climatecommission.gov.au/topics/the-critical-decade/

Attempting to explain the warming trend since the mid-20th century, scientists have carefully considered solar radiation, multi-decadal or century-scale patterns of natural variability (such as the Medieval Warm Period), and shorter term patterns of variability (such as El Niño-Southern Oscillation and the North Atlantic Oscillation). None of them account for the observational record nearly as well as the anthropogenic theory.[4]

The theory of anthropogenic global warming is corroborated by a very large body of internally consistent observations, experiments, analyses, and physical theory. The evidence keeps accumulating. To cite a recent example, consider the November 24, 2009 report called "The Copenhagen Diagnosis: Updating the World on the Latest Climate Science."[5] 26 veteran climate researchers conclude that several important aspects of climate change are worse than the ranges predicted only a few years ago. Even the worst case scenarios of a few years ago were too optimistic. The report also notes that global warming continues to track early IPCC projections. They say that in the absence of dramatic emissions cuts, global mean warming could reach as high as 7 degrees Celsius by 2100.

The most significant new data about global warming:
- "Satellite and direct measurements now demonstrate that both the Greenland and Antarctic ice-sheets are losing mass and contributing to sea level rise at an increasing rate.
- "Arctic sea-ice has melted far beyond the expectations of climate models. For example, the area of summer sea-ice melt during 2007-2009 was about 40% greater than the average projection from the 2007 IPCC Fourth Assessment Report.
- "Sea level has risen more than 5 centimeters over the past 15 years, about 80% higher than IPCC projections from 2001. Accounting for ice-sheets and glaciers, global sea-level rise may exceed 1 meter by 2100, with a rise of up to 2 meters considered an upper limit by this time. This is much higher than previously projected by the IPCC. Furthermore, beyond 2100, sea level rise of several meters must be expected over the next few centuries.
- In 2008 carbon dioxide emissions from fossil fuels were about 40% higher than in 1990. We are in danger of being unable to limit warming to less than 2 degrees Celsius, which is widely considered a crucial boundary beyond which lie a sharp increase in weather-related catastrophes.

The report concludes that global emissions must peak and then decline rapidly within the next five to ten years for the world to have a reasonable chance of avoiding the most terrible impacts of climate change.

A warmed world would result in steeper costs from a number of mounting hazards. We could see worsened droughts, longer and deadlier heat waves, and more damage from windstorms. Each extra degree in the world's mean temperature increases the risk of rising seas with costly damage to coastal cities, and more flooding in low-lying areas. Changed landscapes might stunt the productivity of farms, impoverish

---

[4] Two important websites do an excellent job of covering these controversies and serving as repositories of expert discussion. One is Real Climate, and the other is Skeptical Science.

[5] Ian Allison et al.. "The Copenhagen Diagnosis: Updating the World on the Latest Climate Science," http://www.copenhagendiagnosis.org/executive_summary.html

biodiversity and cause mass extinctions, as well as widening the geographical ranges of infectious diseases. Harsher stresses on ecosystems could stoke wars and regional conflicts over water and other resources. We would be gambling in a collapsing casino where the odds against us increase, our savings dwindle… and we can't leave.

From bias to denial

Unfortunately, cognitive biases are not the only obstacle to climate action. Politically motivated propaganda also plays a major and malign role.

Attempts to cut the world's carbon emissions since the 1997 Kyoto Protocol have provoked a wave after wave of obfuscation, harassment and disinformation.

Fossil-friendly lobbyists reiterate the same often-debunked arguments against cutting emissions, and never fail to find a credulous reception in media outlets. The industry-funded funded deniers continually raise the same bedraggled theories. They claim that the sun is responsible for the warming, or that temperature trends actually show cooling. They claim that a consensus about global warming doesn't exist among climate scientists, or that the global climate models used by the IPCC are unreliable.

These rickety debating points are refuted by long-suffering climate scientists, burned out by the sure knowledge that in a few months the same points will return, trumpeted as brand new discoveries. The distorted science always finds a respectful hearing in the US, partly because of the media's devotion to "balancing" both sides of a debate, no matter how lopsided, but largely because of the amply funded Denial Machine. In the US, an entire political party is determined to ignore climate realities.

Our economy powers itself through fossil fuels, which possess a number of hard-to-replace advantages, including energy density. To date, renewable energy options cannot be scaled up to replace fossil fuels in a straightforward way. These facts loom large in the envenomed politics of climate change. "The American way of life is not up for negotiation!" is an applause line that many avail themselves of. Political and business leaders show a ghastly determination to steer us all toward runaway emissions and escalating catastrophic impacts for the sake of temporarily maintaining our current level of consumption.

The denialist attack is often directed at the scientists themselves. This intensified in November 2009 with the theft and publication of decades of e-mail among climate scientists from servers at the University of East Anglia in the UK. The e-mails reveal some scientists' expression of anger about the denialists, and about slipshod peer review at some scientific journals that published marginal denialist papers.

The denial machine has seized on specific word choices in hastily written e-mails, subjecting them to deliberate misreading. Lost in the media flurry is the most salient fact that none of the bad-temper and sloppy writing in the e-mails in any way alters the climate science. Vicky Pope, head of the UK Met Office (in charge of Britain's climate research) stated tartly: "This is a shallow and transparent attempt to discredit the robust science undertaken by some of the world's most respected scientists."

Dust-ups like the e-mail theft are a potent distraction. Researchers have been forced to devote valuable time to defending their disciplines or themselves. The right wing continues to insist that the e-mails prove the science a hoax. For example, when the Copenhagen Diagnosis was released, the scientists involved had to spend most of their time answering questions about these stolen e-mails. US politicians have been undertaking personal attacks on climate scientists in West Virginia. Academic researchers in the field have been getting death threats in Australia and elsewhere.

A precedent for this exists. The tobacco industry's success in blurring the public health research about smoking worked through a similar strategy. As Naomi Oreskes, a historian of science, has pointed out, this similarity isn't accidental – some of the very same people involved in denying the link between smoking and health trouble are denying the link between human carbon emissions and climate change[6]. Other historians have noted that not even the tobacco industry attacked the integrity of entire fields of science.

Can we avoid a toxic repeat of this decades-long obstruction? Wealthy companies despise talk of anthropogenic global warming because they are convinced it threatens their profits. They believe, albeit mistakenly, that they cannot prosper in an economy that regulates and tries to limit carbon emissions. Thus, their strategy is to fund the denial business, particularly in the US. The media echo this agenda, sometimes because the denialists are advertisers, sometimes because they don't want to draw conservative attack. According to this conservative world view, climate scientists are besieging mainstream values.

The denial campaign has been working. In 2009, the Pew Research Center polled 1,500 Americans about climate change, and 57 per cent agreed that solid scientific evidence demonstrates that the globe is warming. However, this number has declined from 77 per cent in 2007.

Against such an opponent, appeals to science, reason, or even basic prudence have so far been pointless with people who are deeply suspicious about science. The foundation of trust that used to exist for scientific information has been badly eroded. Words like "comity," "ethos," or "mutual cooperation" have an annoying, sentimental air to the ears of a huge audience that would rather not hear about climate risks.

To think and act on a global scale takes an ethos of cooperation and trust. It's not a matter of forcing everyone to believe the experts – but the experts' contribution must enter into the political discussion. The culture war compromises that resource.

---

[6] Naomi Oreskes and Erik M.M. Conway, *Merchants of Doubt: How a Handful of Scientists Obscured the Truth on Issues from Tobacco Smoke to Global Warming*, Bloomsbury Press, 2010, 2011 (reprint).

Risk management

What can risk management do in such toxic, polarized circumstances? Since insurers are prone to the same blind spots, savvy risk managers will continue to do what they've always done, using diversification and risk-spreading to improve social and/or private welfare in conditions where uncertainty abounds[7].

Diversification cannot eliminate all risks, of course, but the power of insurance to spread risk can make a big difference. The insurance industry parcels out the risks that cannot be eliminated across a wider pool of customers. In this way, no one person or small groups of people face overwhelming losses alone.

Insurance is a time-tested method for handling uncertainties with maximum flexibility. It's not ideal for any single outcome, but it works well enough across a broad range of scenarios. Where the actual odds are unknown, the world relies on insurance to deal with low-probability but high-cost catastrophes. Risk managers routinely allocate some resources to reducing the probability or harm of such catastrophic risks. We can argue about how much we should spend, raising the amount if we're cautious, lowering it if we're not -- but we know it makes sense to hedge.

This struggle to judge the risk of large catastrophes exists in almost all kinds of insurance, even where climate change is not an issue. If a danger is large enough, it makes sense to spend some money on insurance, even if the odds are very low.

Insurers want certainty, even when they know that certainty is unattainable. At a conference about hurricane science for an insurance audience, the world's top climatologists discussed modeling and hurricanes[8]. The head of underwriting at a major North American insurer listened to the presentations, impatient at the hedged, qualified way the scientists state their conclusions. The underwriter then complained, "Why don't the scientists give us numbers we can use! These probabilities are too nebulous for us to write business with them!" As far as he was concerned, he was not getting the help he needed from the scientists.

This need explains and justifies reinsurance. Reinsurers serve the same risk spreading needs that insurance companies do, only their clients are insurance companies themselves. Reinsurers focus their attention in part on longer range, more fundamental problems, and survey a variety of hazards globally. Instead of thinking one year ahead, they try to think three years ahead. They warn their clients about emerging risks, one of which is climate change. Others include nanotechnology, parts outsourcing, antibiotic resistance, pervasive computing, and problems with genetically modified organisms. They are the sentinels watching for emerging risks that most people and most businesses haven't started worrying about yet.

---

[7] Gary Yohe and Robin Leichenko, "Chapter 2: Risk Assessment and Management," Climate Change Adaptation in New York City: Building a Risk Management Response, NY Academy of Sciences (Wiley-Blackwell), 2010, page cit. to com.
[8] The 2006 RMS Hurricane Eyewall Symposium, (New York, October 12, 2006).

Events have more than one cause. Some of the other factors that play a large role in economic losses from exteme weather are urbanization and the build-up of value at risk. Many more of the earth's inhabitants (over half by some counts) are living in cities today which means concentrated vulnerability to natural disasters. In hazard-prone areas, this urbanization and increase of population translate into increased concentration of exposure. For an event like a hurricane, global warming might only be, say, a two percent part of the overall risk. If the climate change contribution grows from two percent to five percent, while it seems negligible, in fact it's quite significant. What's more, even the most astute insurer cannot assign an exact number to the role played by climate change in an increasing risk.

The same pattern is true for other impacts associated with climate change. There have always been floods, extreme weather, and periods when the water cycle intensifies. But if climate change is turning up the dial, these familiar events may break out of their usual bounds and become more frequent, or more intense, or just change in unexpected ways. What's more, climate instability tends to make floods, windstorms, and other extreme weather more interrelated.

We can illustrate these tangled questions of causality with a story about Kerry Emanuel, a climate physicist who has argued that climate change is making hurricanes more powerful, though not more numerous. At an insurance event in 2006,[9] a person in the audience asked him about an editorial in the Wall Street Journal, which pilloried environmentalist alarm about climate change and declared that climate change had nothing to do with Hurricane Katrina. To the astonishment of the audience, Emanuel said, "Actually, I agree with a lot of what that editorial says. I don't believe that climate change had anything to do with Katrina."

He noted that in 1965, Hurricane Betsy badly damaged the Bahamas and Louisiana, including terrible floods in New Orleans. He and a number of other then graduate students who are now the venerable pillars of climatology signed a petition saying, New Orleans is a disaster waiting to happen again because of subsiding soil and poorly maintained levees. Shoddy human habits and poor land use governance were making the city vulnerable. Emanuel noted that this petition came about seven years before climate scientists began to take note of climate change as a serious issue.

He also reflected that friends and neighbors have said that they used to disagree that climate change exists, but after Katrina they decided that the climate physicist next door was right. Emanuel said, "I'm torn – I'm pleased that people now agree with me, but they've swung over to my side for the wrong reasons."

Cognitive bias in insurance

Politics is often cognitive bias writ large. Risk management can get shoved aside because of politics. This is usually not the fault of insurers themselves. Following are two examples where politically motivated denial of the facts has badly disrupted sound risk management.

---

[9] Kerry A. Emanuel, Ph.D., "Is Global Warming Affecting Hurricanes?," (paper presented at the William J. Parkinson Distinguished Lecture Series at the St. John's University School of Risk Management, Insurance and Actuarial Science, New York, December 13, 2006).

Following bruising hurricane losses in 2005, insurers studied their catastrophe models and other underwriting tools for Florida property and realized that premiums were too low. But Florida politicians capped the premiums in the belief that they were keeping insurance affordable for owners of coastal properties. Rather than risk further open-ended payouts, insurance companies abandoned the market. The Florida legislature responded by creating Citizens Property, a state insurance program for risks that private carriers would not cover. This subsidy shielded Florida property owners from paying to insure for the true risks.

The state of Florida currently runs a $6 billion deficit. Without huge tax increases, it could not satisfy claims in the event of major hurricane losses. Citizens Property has identified $17 billion that could be obtained through assets on hand, loans and backup coverage, but the estimated damage from a major storm to Citizens' policy holders could be on the order of $23 billion or more. The State is now weighing options, including raising the premium on Citizens' property holders by as much as 10% a year, to encourage policy holders to switch to private insurance carriers.

If state plans such as Citizens are under-funded, then taxpayers who do not own at-risk properties in Florida will foot the bill for major catastrophes when the coffers at Citizens Property run dry.

Yet in Florida the law, business practices and general culture are geared to developing every square inch of land near water – oceans, certainly, but also lakes, streams, wetlands. Even in the absence of climate change, this is a plainly dysfunctional policy. It's also very popular. John Coomber, former CEO of Swiss Re, once grumbled that every American wants to live on the most vulnerable beaches they can find in Florida.[10]

Governments occasionally try to buck the pro-development tide, but the political pressure against the anti-development forces is swift and merciless. Certainly no politician can withstand it. Thus the insurance industry is weaker than it appears when in matters of changing social and economic policies. The hard-wired, entrenched land use policies in Florida are much stronger than the risk managers. The only way to change them would be for other social forces to align with the insurance point of view.

Flood insurance in the US is in a similar state.[11] Historically, the insurance industry avoided flood insurance because charging enough premiums to cover potential losses was almost impossible: flood zones are not large enough to spread risk effectively. Under the federal program, homeowners in certain zones are required to buy policies from insurance companies -- about 90 provide it -- and the government pays for flood damage with federal funds collected largely from homeowner premiums (about $2 billion annually). Few carriers provided flood insurance until 1968, when Congress created the National Flood Insurance

---

[10] John Coomber, personal conversation, November 2005
[11] "The National Flood Insurance Program," the New York Times, September 30, 2009.

Program, now operated by the Federal Emergency Management Agency as a public-private backstop for private insurers.

The government -- and therefore taxpayers -- bears all the risk. It is not a vigorous program.   Yet the government is eager to expand flood coverage, especially outside the most hazardous areas of flood plains. Broadening and diversifying the risk pool of customers lowers the government's costs and protects taxpayers. But those homeowners who are high and dry have absolutely no reason to buy insurance.

This entire mindset has the perverse effect of encouraging moral hazard: homeowners who are insulated from paying for the true cost of flooding have an incentive to build on flood plains. Because the government takes on all the risk, individuals keep adding property value in flood-prone areas. An IPCC scientist once described watching a news show in California, in which a photogenic grandmother watched her house borne away by a flooded stream. When the newscaster asked her how she felt, she said that it was terrible, but they would rebuild, just as they have *seven times before*. The scientist said, "That's the problem in a nutshell. In a rational flood regime, that property would have been condemned and the owners somehow compensated. As it is, we've gone on to incur millions of dollars of easily avoided damage."[12]

Anthropogenic climate change is causing the risk landscape to shift, and not in promising ways for owners of coastal or floodplain properties. If flood insurance were priced according to the actual risks, it would be prohibitively expensive and building on flood plains would slow or even stop. Insurance can help reduce risks – by sending a price signal that can change behavior.

Insurance industry prescriptions

Noted analyst Evan Mills explains: "Insurance is a form of adaptive capacity for the impacts of climate change, although the sector itself must adapt in order to remain viable. A recent Zurich Financial report called "The Climate Risk Challenge"[13] examined some of the potential adaptations and made recommendations:

"Terms and conditions of the policy response must continue to allow insurers to use their core skills to send risk-based price signals and manage risks...."

This is crucial. Sometimes insurers send the wrong signal: they may know a great deal about risk, but they are not omniscient. Or they might send the right price signal, but the message is unpalatable. That's what happened in Florida. Ardent free-marketeers start using political muscle when the market signal tells them

---

[12] Personal conversation with Kris Ebi, Executive Director, IPCC Working Group 2 Technical Support Unit - Impacts, Adaptation, and Vulnerability

[13] Zurich Financial Services Group, "The Climate Risk Challenge: The Role of Insurance in Pricing Climate-Related Risks," (Zurich, 2009)

something they do not want to hear. But risk-based pricing is the voice of reality audible through the noise. You might ignore the science in your underwriting, but sooner or later you will pay a price for it.

The Zurich report also recommended: *"Climate policy must close the global governance gap, including provisions which allow for the quick and efficient resolution of situations involving a conflict of laws both within and between sovereign jurisdictions."*

In practice, this would force governments to take the one step that they always avoid, such as halting the build-up of property in vulnerable areas. We have seen several cases of this. In Ventura, California, local leaders decided to let some beach infrastructure continue to erode rather than repair it, only to face another bill resulting from sea level rise.[14] Elsewhere in California, San Francisco planners have urged that zero-development take place in vulnerable areas -- and local real estate stakeholders are pushing back. In a similar vein, local councils in Australia have issued stern rules about building near the beach, which is a major cultural and commercial change for that country. In the years to come, we're sure to see more locations dealing with these issues.

Another example is prestige projects, particularly dams. Although these can have very negative ecological effects, once a government starts a dam project it rarely abandons it. The battles in the US in the 1960s and '70s between the Bureau of Land Management and the environmentalists are being replayed now in China, with the Three Gorges Dam, or in South Korea, where a long-planned, river-wrecking dam project has just gotten underway.

China's dam building plans for the Yangtze and other rivers face additional political problems. Because much of the rest of southeast Asia depends on this water flow, the dams would condemn downstream nations to an even more parched future. This is already a major tension in the region that will only intensify, since China's water insecurity is a major driver for dam construction.

A further Zurich recommendation: *"Climate policy must enable markets to function properly. To do so, public policy makers must properly assign property rights, and where they cannot be assigned because the property in question is a public good, governments must align incentives to reflect the goals of climate policy."*

Easier said than done. Here again, the "aligned policies" will halt development in vulnerable areas, which is anathema to builders and vacation-home buyers. Markets can direct behavior changes, but they probably won't be as fast and decisive as what's needed.

The final Zurich recommendation: *"Climate policy must recognize the regional nature of climate change and the resulting intersection of energy, water, and carbon risk management strategies."*

---

[14] Tony Barboza, "In Ventura, a retreat in the face of a rising sea," Los Angeles Times, January 16, 2011

Coastal regions need different rules than dry, landlocked areas. A number of straightforward ideas exist and are feasible right now: for example, more extensive arrangements for telecommuting could be made, including rules that mandate staying home and not traveling during extreme weather. . People could be rewarded for sticking to stronger, regionally specific building codes, or relocating from vulnerable areas, or mandating steps to handle business interruption. Such measures would take more political will and more long-term thinking than anyone has yet been able to muster.

Insurance can nudge behavior in safer directions, and encouraging individuals and businesses to think more long-term has decided advantages. But this is insufficient on its own to drive the kinds of measures that our struggle with climate change will require. So we should value the industry for the genuine strengths it brings to the battle against climate change, and not demand more from insurers than they can deliver. That impetus will have to come from other sources.

The analogy with financial risk

Financial risk provides a close analogy with climate risk. In both cases, cognitive biases often run rampant, goading participants into taking far greater risks -- and resulting in worse catastrophes. In both cases, iterative risk management has often been overridden by ideological agendas and greed, and major stakeholders oblivious to risks have framed the public debate to their liking and have taken over their regulators.

In both climate change and finance, regulatory capture undermines the effort to grapple in a fundamental way with the specific situations. While large emitters have proved skillful at blunting the effect of regulations to cut emissions, they have much to learn from financial firms: Wall Street, through years of antiregulatory propaganda and bottomless spending on lobbyists, has almost entirely vanquished Washington. Decades of anti-regulatory, pro-market campaigning have successfully narrowed the debate.

In the years preceding the credit crisis, a number of economists and market analysts spoke out against the real estate and financial bubbles. Some commentators pointed out the dangerous consequences of dissolving the Glass-Steagall Act in 1998 (a longstanding barrier separating commercial banking and investment banking), and of failing to regulate derivatives in any way. The financial media paid scant attention to them.

Before the 2008 crash, pro-market figures like Fed Chairman Alan Greenspan squashed attempts to regulate derivatives or otherwise rein in the markets, energetically assisted by Democratic appointees like Robert Rubin and Lawrence Summers. The anti-regulatory ideologues in finance were also the regulators.

Large Wall Street firms and the media companies that serve them portrayed the sturdy regulatory framework that had emerged in the New Deal as a hindrance to financial dynamism. The business and conservative press has worked hard to disrupt the spirit of cooperation and willingness to reform that characterized the New Deal, and after years of effort, its reputation and achievements had been belittled and marginalized.

Meanwhile, much derided "Cassandras" such as economists Nouriel Roubini and Robert Shiller warned against the slackening of mortgage standards and lending discipline. This potential for financial losses became far worse with the introduction of headlong or sometimes deliberately fraudulent securitization.

The changed attitude toward risk involved many major investment banks. Some of them, chastened by large trading losses in the 1980s, had introduced strong risk management programs. When these systems began signaling problems in the 2000s with the new securities, the risk managers were sidelined or dismissed. Basically all of Wall Street, with few exceptions, bundled tranches of very risky loans into new securities.

Those who cautioned about the financial risks were in a similar position to the IPCC, warning that current practices were leading to major risk accumulation. The investing public and regulators were disinclined to disturb the process because everyone was profiting. The usual cognitive biases kicked in: those making money tend to deny, object, and minimize warnings of future or of present trouble. It's hard to be reasonable when everyone around you is profiting. And as with climate, these biases dovetailed with the propaganda of the free marketers.

In a more egregious turn, the ratings agencies, paid by Wall Street, blessed these instruments with high ratings, despite major conflicts of interest. With the marketing of these securities worldwide, an ugly but localized problem became a global catastrophe.

Even after the recent financial turmoil, the same faces and the same figures remain in charge, even though their policies have resulted in mammoth economic and financial losses. The financial markets version of the culture war vociferated against any attempt to regulate the markets, enforce transparency, or punish criminal behavior.

Climate change activists and financial reformers face similar obstacles – most people see current arrangements as a force of nature. In this traumatized post-bubble environment, the only ideas that emerge from conventional sources consist of a handful of minor changes. In finance, many of them boil down to the faith-based desire to let Wall Street regulate itself because government regulation is incompetent and destructive by definition. To reduce carbon emissions, the large nations have proposed minor reductions that fall well short of the effort needed to keep levels of greenhouse gases at safer levels. In both bases, ideology has successfully shoved the risk managers aside.

The big one

The analogy between finance and climate is not exact. Economics is a less developed proto-science compared to the scientific work assembled by the IPCC, in which data from physics, chemistry, oceanography, and numerous other disciplines have converged in a highly corroborated way.

Climate Change Adaptation in 2011

To cite just one example, in economics peer review is hardly as demanding as in climate science. Some market analysts spout errors and false predictions year after year, and yet they prosper. Some journalists and others try to point out the naked emperors, but their debunking has little effect. The entire field is fragmented into camps – a huge difference from climate science, where the consensus is quite strong (despite the pretensions of denialists to the contrary), and errors are systematically weeded out.

The analogy breaks down in another way. Economic and financial problems are more easily fixed than climate change. If the politically tough but appropriate steps were taken by financial regulators, the capital markets and the larger economy would quickly improve. But climate change is a larger and more intractable challenge. If we magically eliminated all greenhouse gas emissions today, the world's climate system would still be committed to several decades of dangerous rises in mean temperature. Even if we take all the right steps, the struggle to slow down global warming is going to last the better part of a century. Reducing emissions demands a coordinated global response that will counteract cognitive biases and formidable ideological enemies. We need a greater sense of urgency about fighting climate change.

We've done it before. Before World War II, isolationist sentiment was strong on the right, and there was also a substantial leftwing disinclination to get into the war. The national ethos was fragmented. Pearl Harbor changed that.

During the five years of the World War II, the US and its allies utterly transformed their economies and their societies in the face of a huge challenge. Tremendous industrial energy and creativity were brought to bear on the logistical challenge of defeating a ruthless, powerful enemy. The Allies even found ways of cooperating with erstwhile enemies like the Soviet Union.

Climate change is global, impersonal, and it draws its terrible force from our own actions – our emissions into the atmosphere. It took Pearl Harbor to silence isolationist sentiment in the US. Will it take the climate equivalent of Pearl Harbor – an event so overwhelming that it switches public opinion in a single day?

Climate in bondage

Attempts to cut the world's carbon emissions since the 1997 Kyoto Protocol have provoked wave after wave of obfuscation, harassment and propaganda. More than ever before, climate scientists are under attack by politicians and other industry voices promoting manufactured pseudo-events like Climategate. Fossil-friendly scientists churn out the same old often-debunked arguments against cutting emissions and never fail to find a credulous reception in media outlets.

In the US, an entire political party is determined to ignore reality about climate in the face of mounting evidence. To today's Republicans, it doesn't matter that an enormous and growing body of internally consistent observations, experiments, analyses, and physical theory points to the increasing atmospheric concentration of greenhouse gases, with carbon dioxide (CO_2) the most important, as the ultimate cause for the observed warming.

Our economy powers itself through fossil fuels, which possess a number of hard-to-replace advantages, including energy density. Renewable energy options can't be scaled up to replace fossil fuels in a straightforward way, at least not to date. These facts polarize the politics of climate change. Talk about conservation, and some prominent Republican will snarl, "The American way of life is not up for negotiation," an applause line that many agree with.

In short, political and business leaders show a ghastly determination to steer us all toward runaway emissions and escalating catastrophic impacts for the sake of temporarily maintaining our current level of consumption.

This moral failing takes several forms. First, the residents of heavy emitting countries spread the impacts of their energy consumption to poorer countries, through outright plunder and through a loss of ecosystem services from wetlands, forests, rivers. The poor in wealthy countries and everyone in developing nations suffer the most.

Another source of moral failing is our relation to the future, as early generations can pass the consequences of their greenhouse gas emissions along to their children and grandchildren. The oppression of later generations stands revealed in the recent discussions of the social discount rate, the view in welfare economics and elsewhere that benefits and costs today are worth more than effects further in the future.

We can discount the more remote effects of our acts and policies at some percentage rate per year. A high discount rate means that we're giving less weight to future generations. A lower rate means we're giving more weight to the future. But in climate terms, the present moral importance of future events does not decline at a rate of x per cent per year. Remoteness in time has no more dampening effect on moral significance than spatial distance.

In grappling with these injustices, we lack even basic conceptual tools. Economists tend to dominate the discussion of climate policy, although in many ways economists are part of the problem. Despite the often observed inability of cost benefit analysis to get a grip on ecological or boundary-crossing moral and political problems, it's still the dominant method for discussion of many climate change policy issues.

As Stephen Gardiner says in his excellent book on the subject, *The Perfect Moral Storm: The Ethical Tragedy of Climate Change*, we don't have strong theories to guide us through the problems posed by "intergenerational ethics, international justice, scientific uncertainty, the human relationship to animals and the rest of nature." Philosophers, let alone politicians, flounder ineffectually with these injustices, which interact and create a problem far more pressing than any single failing by itself.

I suggest a historical analogy that may contain a glint of hope.

Slavery was utterly entrenched in the United States, much the way fossil fuel usage is entrenched in the modern world. During the 19th century and before, many great fortunes rested on the backs of slaves, and much of the nation's economy depended on bonded labor. Certainly the wealth of many powerful people flowed from trafficking in human beings. Slave owners considered themselves upright people entitled to make a living. Even among abolitionists, white supremacy was a common assumption, and many of the arguments against slavery were based on a belief that slaves were racially inferior.

Today, American racism often resurges strongly, as is evident in reactions to the election of Barack Obama, such as birtherism and the routine beliefs of Tea Party republicans. Slavery and its consequences still blight our history and our present, evident in many demographic measures of well-being. Nonetheless, the old racist virulence has been diluted in the decades since the 1860s. Today's well-taken repugnance to slavery today is a distinct moral advance over the 19th century American attitude.

Might something similar happen with climate change?

The forces arrayed against responsible action to cut carbon emissions bear a strong resemblance to the slave owners who fought every attempt to end the human ownership of other human beings. To abolitionists at the time, the slaveholder appeared to hold all the power. But their ascendancy didn't last. Like slave owners in the past, the west has proved adept and ideologically committed to preserving their entrenched system. They shouted down and politically outmaneuvered their enemies at every turn.

Awareness of the moral dimension to climate change is in a larval stage   for many Americans, climate change doesn't have any moral implications at all. The analogous year is 1855, perhaps. Who knows what might galvanize the public? We're still waiting for the latter-day Harriet Beecher Stowe to write a book that will make people see the system they inhabit. Or perhaps some sequence of disasters will persuade people, the way the terrible floods of 1953 persuaded the Netherlands to defend themselves much more thoroughly from the sea. An event that large might mean that we begin negotiating about the American way of life, and that the fossil fuel lobbying will become a public embarrassment.

We just need the transition to a moral outlook to happen much faster.

What lasts?

At the heart of sustainability lies value that endures and grows from one generation to the next. Genuine sustainability rests on three pillars of support – economic, environmental, and social.

- Economically, we seek steadily growing revenues and profits. Our businesses should be self-sustaining, profitable and leave something for those who follow us.
- Environmentally, ecosystems must remain healthy and not degrade from the stress of our activities. To create ecological deficits is to impoverish our children.
- Socially, long-lasting solutions enhance well-being and allow grievances to find orderly resolution. Healthy, sustainable societies have mechanisms for arriving at equitable outcomes to disagreements.

Not all these pillars are created equal. Economic sustainability is the central one and a necessary condition for the other two: without it, nothing lasts for long. At the same time, a single pillar cannot support an entire society, so economic strength is not a sufficient condition for genuine sustainability.

This paradox often clouds discussions of what lasts. Although economic conditions bear the most weight in the short term, genuine sustainability is comprehensive and holistic. The central pillar requires the others for support.

The advent of genuine sustainability

Each stakeholder group possesses both strengths and weaknesses that flow out of the goals it pursues, and what it values most. In addition, each needs to develop its least evolved skills. Everyone's skills will be needed to address the urgent climate and environmental emergency the world faces.

Business

In ways that are immediately apparent, businesses have the easiest task of all stakeholders in pursuing sustainability, since their area of strength is the central pillar. They are already focused on economic factors, in the form of maximizing profits and increasing returns to their shareholders.

This focus often obscures the other pillars. Historically, businesses have tended to overlook environment and society. To many CEOs issues of justice and human rights have been deemed secondary, even when their own supply chains are implicated in child labor, for example, or unsafe practices. They have been slow to see the value of protecting the watersheds where their factories are located.

Businesses that fail to take account of social or environmental sustainability can even jeopardize their ability to maintain their operations and profitability. Neglecting the dimensions of genuine sustainability opens the door to potentially catastrophic reputational risks.

Perhaps as a result, a growing number of CEOs have come to know quite well that an exclusive focus on economic and financial sustainability can endanger their bottom line: they are learning to account for their own carbon emissions, for example, and to verify that links in their supply chains are not violating human rights or child labor statutes. Many managements and investors now track environmental and social indicators as well as financial performance.

Some executives are finding ways to monetize other dimensions of sustainability. They are creating new green businesses which can make direct contributions to their bottom lines, whether via green products, or innovative services that help monitor or reduce carbon emissions.

Governments

Governments at all levels and of all kinds are custodians of social sustainability in the form of public order and to some extent environmental goods. Ideally, they create a society where all citizens have input into their own destinies. They intend to be known as inclusive and unprejudiced.

The economic pillar for governments needs reinforcing. Even though they have been made painfully aware that taxation is often not an economically sustainable option, since push-back from citizens or economic downturns can result in a loss of revenue at vulnerable moments, usually it requires an emergency or a crisis to force governments to heed economic concerns.

From the standpoint of genuine sustainability, the challenge for governments, to enhance the quality of life for their citizens, is to move away from a static regulatory model, and towards an emphasis on their own competitiveness, and even towards revenue and profitability. They need more economic skills as they strive to promote economic development in their specific jurisdictions.

Some governments have begun to engage more energetically with economic issues: several, for instance, have learned to appreciate that ecosystems within their borders can be a magnet for tourism and economic growth. They have put sophistication and effort into their own economic development. In Costa Rica, for example, much of the country is a protected area, and a major source of revenue from foreign visitors. Since the land is protected, the nation gets the full benefit of the ecosystem services provided by pristine forests and rich biodiversity. It's an ecological plus that directly creates economic value.

NGOs

Nongovernmental organizations are harder to categorize, but their mission often involves easing environmental and social harms. NGOs are born because of ecological or social ills that the other stakeholders are neglecting.

Many NGOs maintain that that the effort of assuring a flow of donations takes so much time that emphasizing economic sustainability can seem like a further distraction from their mission. By temperament and culture, many NGOs prefer to operate in a marginal way rather than make their economic base stronger. This limits their effectiveness.

NGOs that want to enhance their global relevance, expand their donor base, and play new mediating roles between citizens, businesses and governments are beginning to recognize that promoting their social or environmental agenda often requires them to integrate economic and financial considerations in new ways.

Complexity creates risk

Promoting genuine sustainability involves daunting uncertainties.

It's often hard to find the best route to make economic sustainability primary and at the same time remain mindful of environmental and social issues.

Economic turmoil has harrowed our systems over the last two years, with no outcome in sight. It's unclear who is capable of restoring the world's financial systems in a lasting way, or how it will be done. Macroeconomic risk shows little sign of abating. The primary pillar of sustainability still needs a firmer foundation.

In addition, the world economy suffers from a longer-term economic uncertainty, one that directly involves the other dimensions of sustainability. No game-changing technology has emerged to solve our abiding energy and environmental bind. The transition from an energy system reliant on fossil fuels to renewable might take fifty years, according to energy economists. Such transformations are rarely linear and orderly. Nor is there a consensus on how to best manage our remaining fossil fuel supplies. This global economic and environmental sustainability issue poses great risks – to cite just one, the danger of backing the wrong technology.

A related major unknown is global warming. Climate change has been described as a slow-moving emergency, exactly the sort of risk that humans and human societies find difficult to manage. The overall shape of climate impacts is well corroborated, but the local consequences are much harder to predict. Adapting, even enduring, will be expensive, perhaps ruinous.

Social factors further complicate this picture. The least developed countries are already suffering the most from climate instability, and yet their historical carbon emissions are negligible. The knotty questions of justice and fairness this raises make a climate agreement difficult to reach. On a global scale, climate change presents social and environmental risks. A weak economy worsens already existing social ills, from poverty to civil rights.

Faced with a phalanx of colossal challenges, we need to find the right steps in the right order. We contend that the first step must be economic.

Troubleshooting and trouble sharing

Public/private partnerships are a plausible way forward. Such coalitions become more attractive as issues loom too large for a single stakeholder to address alone. The primacy of economic sustainability also gives governments and NGOs compelling reason to seek partners with economic skills.

We are already seeing NGOs making common cause with individual companies and entire industries undertaking to promote social and environmental goals. Cities and states have made great strides in allying themselves with businesses to promote economic development. In the pursuit of genuine sustainability, such connections are often expressed in new and unexpected forms, such as the transformation of Essen's blighted industrial landscapes into beautiful, imaginative parks. These hybrid partnerships may take new shapes that will emerge organically, and probably cannot be legislated in advance.

If handled well, such alliances reduce risk and uncertainty for all players. They can provide profits to businesses, as well as unexpected revenues to governments and NGOs. They allow each participating stakeholder to complement the others. Most importantly, they promote economic sustainability, which supports everything else.

Making effective partnerships demands flexibility from the participants. They must first be able to identify likely candidates, and then teach themselves how to work effectively together. Each one must learn enough of the other's language to communicate effectively. (Many NGOs, for example, feel put off by business's exclusive focus on profitability. Business executives may find that NGOs and governments appear to ignore economic consequences.)

The major trends

In a landscape of multiple uncertainties, some global trends stand out as likely candidates for inventive partnerships.

Water scarcity

Climate change dramatically affects water availability. It can cause sea level rise, a major factor in water stress that is resulting in saltwater intrusion in Vietnam and other coastal countries. It also causes an intensified water cycle that can result in desertification in many locations. Peru and other mountainous areas are often the most vulnerable to these effects. In the American west, reduced flow and mounting silt threaten the ability of the dams on the Colorado River to generate electricity. This imperils the economic underpinnings of a large part of the country – including California.

Many water sources, in use for decades or even centuries, are now being compromised. Rivers around the world are overdrawn. Some locations with nascent environmental laws (in Indonesia, India and elsewhere) are over-pumping aquifers. The economic consequences of such overuse are already significant.

Agriculture is responsible for some 70 percent of water withdrawals, and prolonged water scarcity can erode the economic basis of farming. Irrigation practices can have major pollution consequences. Water stress awakens land use controversies that involve other aspects of genuine sustainability. Water problems threaten food security, and this will be a major concern in the years ahead.

Water shortages also provoke sharp social quandaries about fairness and access. As water dwindles, governments need to find the most socially equitable way to handle its distribution.

Large scale urbanization

The rapid growth of cities has become an immense and dominating feature of modern life, as once-small towns rapidly boom into metropolitan regions – such as Shenzhen in China, or Lagos in Nigeria, or the growth of Las Vegas that preceded the collapse in real estate.

In specific ways, cities are ideally positioned to promote genuine sustainability, although the process will hardly be predictable, or smooth. The crucible of urbanization also shows how economic power sets the boundaries for other kinds of sustainability.

Huge slums and favelas in developing countries may seem like magnets for social ills. But as Stewart Brand (of *Whole Earth Catalog* fame, and the author of the more recent *Whole Earth Discipline*) and many others point out, they are also economic dynamos. Brand points out that city dwellers have a much smaller ecological footprint than rural dwellers, using less energy and making more efficient use of resources. Moreover, cities are "…where vast numbers of humans … are doing urban stuff in new and amazing ways. People are trying desperately to get out of poverty, so there's a lot of creativity … The United Nations did extensive field research and flipped from seeing squatter cities as the world's great problem to realizing these slums are actually the world's great solution to poverty."

Cities face ordeals in governance and fairness in their attempts to enfranchise all their citizens. City governments struggle to maintain the economic energy of their slums while making them more livable for inhabitants.

As a bold plan for how cities might develop in the future, consider the green city planned for Masdar in Abu Dhabi, which will draw electricity only from the sun and other renewable sources. A zero-carbon, zero-waste infrastructure will be built in from the inception. Covering a little over two square miles, Masdar will house 45,000 to 50,000 people and 1,500 businesses, primarily commercial and manufacturing facilities for products. Cars are banned, with everyone traveling via public mass transit and personal rapid transit systems. As an attempt to build a genuinely sustainability city, Masdar is example that all stakeholders will study for years to come.

Increasingly mobile networked society

Information and computing technology (ICT) creates the possibility of a "virtualized" workforce. People can work productively with minimal commuting, with less strain on roads, improved air quality, and eased fossil fuel consumption. Access to the network can have educational pluses and increased domestic benefits. Businesses and cities that take advantage of these new possibilities will experience enormous boosts to their economic well-being, and that helps both environmentally and socially.

Investing in mobility through ICT is one way countries can support and reinforce the economic pillar of sustainability. Such investment presents enormous business opportunities to industry, municipalities, education systems and local administrations.

South Korea, for example, has invested heavily in broadband capability and as a result leads the world in broadband penetration. In addition to providing tremendous economic benefits, this also is resulting in new ways of working and interacting. The Korean government has put much of its activity online, so a transaction that used to require time-consuming personal odysseys can now be accomplished with a few clicks. Other governments have used blogs, surveys, and social networking to make themselves more available. People like their government far more when their experience involves a quick on-line solution to an issue, rather than a lost day waiting in line at an office.

In developing countries, the mobile phone has become the engine for startling innovations, enabling a nation to bypass the legacy technology of telephone lines and go straight to the most sophisticated mobile services.

Technology-enabled changes are underway in promoting e-health, digitizing medical records and empowering telemedicine around the world. Individuals can now access very sophisticated medical information, which spurs medical professionals to keep abreast and learn how to communicate with a digitally informed audience.

A broadband-intensive mobile economy raises several old issues in new forms. Privacy breaches in a networked society can do terrible economic and social harm. The cost for amassing highly valuable data on millions of citizens keeps dropping, and the ease of data theft keeps growing. It's a recurring story – a single worker at a credit card company has their laptop stolen, and suddenly the personal data of millions of people is compromised, with enormous resulting costs.

ICT requires abundant, steady electrical power, and a country can sacrifice environmental value if they are drawing power from unsustainable sources. If they rely on coal, they also incur major public health deficits. Thus a networked society instantly puts pressure on environmental sustainability.

Network innovations can emerge anywhere, and developing countries are often the most creative at finding solutions to persisting problems. Just look at M-PESA (M for mobile, pesa is Swahili for money). A mobile-phone based money transfer service that was born in Kenya, M-PESA is a branchless banking service that enables users to deposit, withdraw and transfer money anywhere, pay bills, and buy airtime for their phones. Its spectacular success in Kenya is spreading internationally.

Crumbling or outmoded infrastructure
Cities and countries face huge outlays as a result of decades of neglect and deferred maintenance. Roads, bridges, dams, and other fixtures of the built environment, particularly in the US, have steadily eroded. A recent report by the US Engineering Society gave bridges in the United States a grade of D+. Some economists have described inadequate infrastructure as a hidden tax, and a definite obstacle to economic sustainability.

Because of climate change impacts such as sea level rise and flooding, straightforward repair often will not suffice. Rising waters are likely to be the most expensive climate change impact for many countries, and the best adaptation strategy is rarely obvious, or palatable. Governments must choose some combination of armoring coastlines, surrendering vulnerable areas to the waves and saving others – or retreating. The large redesigns and overhauls involved in each choice will be costly.

Large coastal river cities, such as New York, or London, or Jakarta, face inundation from sea level rise as well as from their rivers and tributaries. More intense rainfalls, for example, can clog drains and culverts, soaking once-dry areas beyond official flood zones. Subway systems are particularly vulnerable to flooding – as New Yorkers and Londoners can attest.

Land use and development regulations are often outmoded. Devised in a different era, they can be obstacles to imaginative solutions. For example, many cities are changing the rules about brownfield development, enabling old industrial sites to be repurposed without expensive remediation, such as the Essen parks, in which old lime kilns have become planetariums and conveyor belts turn into aerial walkways.

Many urban locations are making a growing case for greenbelts and blue belts as a crucial natural infrastructure, taking an ecosystem services approach to flooding, wastewater management, and pollution control. New York City's "blue field" program, for example, is aimed at preserving existing small wetlands throughout the metropolitan area. Alongside the ecological benefits, esthetics and economics are an explicit goal – a healthy wetlands is an intangible enhancement to the attractiveness of a place.

Business and NGO participation can improve the quality and fitness of infrastructure decisions. They can strengthen the central economic pillar, or they can weaken it. The stakes are high, because smart and collaborative decision making can build structures that brighten – or blight – our lives for generations.

In a way, cities are a microcosm of the world's overall climate situation: the remedial efforts we make now may seem expensive, yet they are far less costly than the ultimate bill that will come due. For cities, keeping bridges, roads and basic infrastructure in good repair can postpone their replacement for a long time. For the world at large, spending 1% of GDP on greenhouse gas mitigation today could save us far more onerous climate change adaptation costs in a few decades.

Genuine sustainability as a paradigm shift

Even in the absence of climate change or concerns about energy efficiency, cities must address flooding and infrastructure repair. Businesses will face disruptions and relocation decisions. Economic decisions will generate social and environmental consequences that NGOs will struggle to deal with.

The major trends also suggest that genuine sustainability is a way of reframing already existing perils. This is an exciting prospect for governments and NGOs because it raises the hope that success in one struggle can encourage success in others. This can be done by first improving economic foundations, so that longer-term the goals of environmental and social sustainability can rest on firm ground.

One recent instance is the suggestion that reducing the inequalities between rich and poor nations could be the main driver for avoiding the worst effects of climate change and even reducing atmospheric levels of the greenhouse gas carbon dioxide (according to UK researchers writing in the International Journal of Global Warming). Studies have shown that economic wealth is the most significant driver of greenhouse gas emissions, far more than population growth, and that a reduced gap between rich and poor can therefore cut down on emissions overall.

In a political environment that is both 100% wired and more democratized, top-down methods are less possible, and certainly not likely to be sustainable. Even in more centrally governed societies, leaders have to take more account of stakeholders' needs and demands. Genuine sustainability probably has a bias towards democracy in the sense of more participatory deliberation, as well as towards economic growth from smaller, local projects rather than huge centralized ones.

Economist Robert Shiller declares that many of the world economy's recent problems result from financial and economic innovations that have not spread widely enough. He says, "We need to extend finance beyond our major financial capitals to the rest of the world. We need to extend the domain of finance beyond that of physical capital to human capital, and to cover the risks that really matter in our lives. Fortunately, the principles of financial management can now be expanded to include society as a whole." He argues that "democratizing means drastically expanding the financial sector so that it plays a deep role in our lives. Such an enterprise would generate whole new industries and major tasks for people to complete." This expansion of finance could play a large role in the growth of genuine sustainability.

The best course for each stakeholder will emerge slowly. Everyone's responses will be diffuse and vague for long stretches. It's difficult to make economics primary and still have resources for the other dimensions of sustainability, but the rewards will be enormous for those cities, governments, businesses and NGOs that make astute decisions in handling these uncertainties.

Under these circumstances, communications grows in strategic importance. Stakeholders must be able to explain their positions in a transparent, persuasive way. They also need to listen. Non-transparent, agenda-driven communications can actually do harm to an organization. Communicating effectively can help all three pillars stand firm.

Five takeaways

•   Sustainability is bigger than climate change and renewable energy – it's a comprehensive approach to economic, environmental and social woes. Economic sustainability is the central pillar of the three. Without it, nothing can last.

•   Whether you're a government, a city, an NGO, or a business, you're already involved in the pursuit of genuine sustainability, perhaps unwittingly, as you're trying to create lasting value, economically, environmentally, and socially.

•   Promoting genuine sustainability often means finding the right set of partners, ones that may not look like you and don't operate the way you do.

•   Genuine sustainability takes time and patience – quick fixes lie elsewhere.

•   To create lasting value, you probably have to live with an uncomfortable degree of uncertainty. The first course of action you choose may need replacing.

Blog Posts by (or About) Brian Thomas

Thursday, July 28, 2011

Adaptation, justice and morality in a warming world

Jeremy Hance of Mongabay interviews your faithful blogger today. I'm basking in the fame!: If last year was the first in which climate change impacts became apparent worldwide—unprecedented drought and fires in Russia, megaflood in Pakistan, record drought in the Amazon, deadly floods in South America, plus record highs all over the place—this may be the year in which the American public sees climate change as no longer distant and abstract, but happening at home. With burning across the southwest, record drought in Texas, major flooding in the Midwest, heatwaves everywhere, it's becoming harder and harder to ignore the obvious. Climate change consultant and blogger, Brian Thomas, says these patterns are pushing 'prominent scientists' to state "more explicitly that the pattern we're seeing today shows a definite climate change link," but that it may not yet change the public perception in the US.

Thomas, who writes frequently about climate adaptation and justice, says that some governments—local and national—are beginning to act on adapting to a new, warmer, and more unpredictable world. However, many are not moving quickly enough.

"A great many coastal towns and cities are acting as if they have centuries to deal with sea level rise. Because the impacts feel like they are a long way off, most people are procrastinating. It's a cognitive problem we face because of the very long-term nature of climate change impacts. The risks are grave, the impacts are here, but the problem doesn't feel urgent," he told mongabay.com in an interview....

Thursday, August 11, 2011

That global threat

Reporter Jack Coraggio of the Litchfield County Times interviewed me yesterday, and the article is already out:...Microcosmically, these represent the costly consequences of a rapidly changing climate, a scientific phenomenon the Intergovernmental Panel on Climate Change [IPCC] refers to as "unequivocal." Amid the overwhelming, sometimes doomsday-like predictions that face us, we can take some comfort knowing that forward-thinking solutions for the management and mitigation of the impact exist.

West Cornwall resident Brian Thomas, an independent sustainability consultant, a new kind of green job, has ideas that are in demand. Already, the Brown University graduate has developed green-themed projects for clients that include Merrill Lynch and the cities of Chicago and New York.

"Science tells us things we don't always want to hear, and it's a challenge to know what to do with it," said Mr. Thomas, currently a member of the New York City Panel on Climate Change, EnviroComm, the Association of Green Technology Auditors and the West Cornwall Conservation Commission. "People don't like to deal with risk, but risk is at the heart of all this."...

Friday, August 26, 2011

Don't blow me down! The sculptures of Tim Prentice menaced by the hurricane

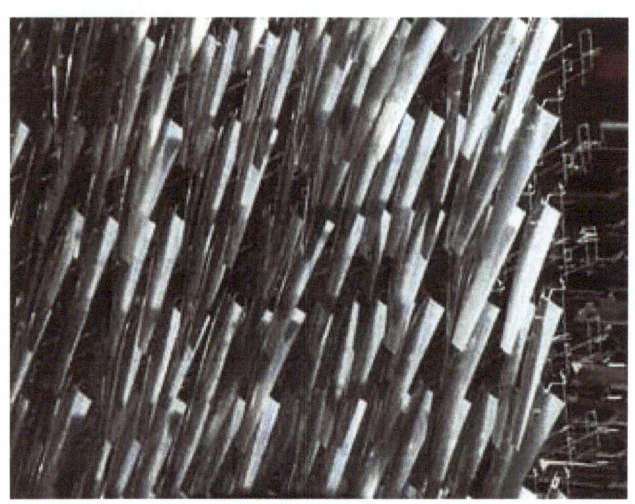

On the Friday before Hurricane Irene reaches the northwest corner of Connecticut, I paid a call on Tim Prentice, a neighbor who is best known as a kinetic sculptor. Tim and his team create mobiles. These delicate structures of wire, Lexan and other materials hover in the air, or are attached to vertical surfaces, where they spin or wave in the breeze. Some move slowly, others are more intricate. Tim's house and shop are a compelling attraction, since his yard teems with startling mobiles unlike anything you'd expect to see on a road in rural Connecticut.

His work has been installed in hundreds of venues, from embassies to corporate headquarters in the United States and around the world. On his website he says, " I take it as an article of faith that the air around us moves in ways which are organic, whimsical, and unpredictable. I therefore assume that if I were to abdicate the design to the wind, the work would take on these same qualities. The engineer in me wants to minimize friction and inertia to make the air visible. The architect studies matters of scale and proportion. The navigator and sailor want to know the strength and direction of the wind. The artist wants to understand its changing shape...."

With nonstop hurricane porn blaring from the media, I thought, how are Tim's fragile-seeming sculptures going to fare in a high wind? I paid a call on him to find out. I found Tim in his barn wrapping a video shoot for a presentation. I asked him whether he was going to bring any of his pieces inside to protect them.

"I don't know," he said. "I'm curious to see how they perform in a high wind. I had one commission that was out in the open, and the client called me before a nor'easter wanting to know if it would hold up. I told them to leave it out so we could see how it would do. But they chickened out and brought it inside."

Tim noted that some of his pieces could handle a 70-mile-per-hour wind, but was dubious about anythng stronger. He never takes the outdoor sculptures inside. "Sometimes after a tough winter we find squares of Lexan lying on the ground, and we have to fix them back up a little bit." With a shrug, he added, "I don't have to decide today. We'll see what the weatherman is saying tomorrow."

As his wife Marie Prentice (a noted poet in her own right) pointed out, Tim was philosophical about the fate of his sculptures because his shop is right there, and he can fix anything that gets too badly chewed up.

The fate of art subjected to climate change impacts rarely comes up in your average climate discussion. But maybe it should, especially when the art exists to make use of the wind, the way Tim's art does.

All photographs by Brian Thomas, which you can use under the Creative Commons license of your choice

Saturday, August 27, 2011

Our former headquarters rather vulnerable

This macho aircraft you see on the deck of the Intrepid Sea-Air-Space Museum is McDonnell F3H Demon, a few blocks from the erstwhile Manhattan branch of Carbon Based, otherwise known as my apartment. We lived on the second floor.

With climate risk very much on my mind, I would tell my wife that a bad hurricane on a dangerous track could send a storm surge higher than our second floor window. We were in flood zone one (I'm not sure whether that's Zone A, the mandatory evacuation zone for Hurricane Irene). She would smile pleasantly at me and treat my dire warning with irony.

Now that we've let that apartment go and are living in the woods full time, she's absorbed in nonstop reports of Gotham inundation, storm surges, flying debris, and other cheery topics on the Weather Channel, which, as Jim Gaffigan insists, is clearly pro-hurricane.

Of course, the actual impact of Irene may turn out otherwise, a rain not even bad enough to disrupt the Falun Gong protests always underway, since our building was across the street from the Chinese consulate. The bad-tempered seagulls that roost on the hawser chains of the Intrepid may just crouch and ride out the wind. We'll see.

Sunday, August 28, 2011

Tim Prentice update

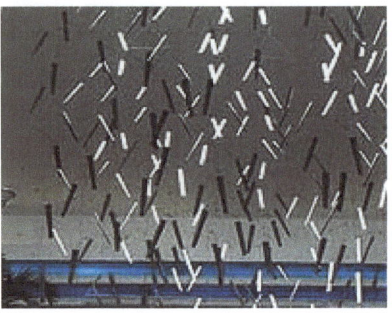

At a town event yesterday, the rain had already started. Tim Prentice sidled over to me and said, "After we talked I took a few pieces in."

"All of them?" His grounds sport some pieces that are twenty feet high.

"No, just the ones that could be readily moved, and the more fragile ones."

"So in the act of reporting the story, I changed the story."

"Yes."

As a journalist, I was disturbed at having intruded on reality like that. But as someone who believes in prudent risk management, I was glad to see some beautiful kinetic sculpture protected, at least for a while.

Shown above, a piece of Tim's called "Flashdance," which installed in Jacksonville, Florida, which ordinarily is a lot more hurricane-prone than Cornwall, Connecticut. I filched the image from Tim's website

Friday, September 23, 2011

Ominous skies

Driving around yesterday with a camera, I saw this sky over a field in Canaan, Connecticut. It reminded me of one of John Ruskin's nutty but prophetic essays, *The Storm Cloud of the Nineteenth Century.* Ruskin noted, "In many of the reports given by the daily press, my assertion of radical change, during recent years, in weather aspect was scouted as imaginary, or insane. I am indeed, every day of my yet spared life, more and more grateful that my mind is capable of imaginative vision, and liable to the noble dangers of delusion which separate the speculative intellect of humanity from the dreamless instinct of brutes: but I have been able, during all active work, to use or refuse my power of contemplative imagination, with as easy command of it as a physicist's of his telescope: the times of morbid [vision] are just as easily distinguished by me from those of healthy vision, as by men of ordinary faculty, dream from waking; nor is there a single fact stated in the following pages which I have not verified with a chemist's analysis, and a geometer's precision."

He also says, in a letter to the editor of *the* St. James's Gazette: "I have been a very constant though not a scientific observer of the sky for a period of forty years; and I confess to a certain feeling of astonishment at the way in which the "recent celestial phenomena" seem to have taken the whole body of scientific observers by surprise. It would even appear that something like these extraordinary sunsets was necessary to call the attention of such observers to what has long been a source of perplexity to a variety of common folk, like sailors, farmers, and fishermen. But to such people the look of the weather, and what comes of that look, is of far more consequence than the exact amount of ozone or the depth or width of a band of the spectrum.

'Now, to all such observers, including myself, it has been plain that of late neither the look of the sky nor the character of the weather has been, as we should say, what it used to be; and those whose eyes were strong enough to look now and then toward the sun have noticed a very marked increase of what some would call a watery look about him, which might perhaps be better expressed as a white sheen or glare, at times developing into solar halo or mock suns, as noted in your paper of the 2d of October last year.

He goes on to blame industrialization for the spread of the Storm Cloud, or "plague cloud" as he sometimes calls it.

Maybe he was on to something.

Saturday, September 24, 2011

What happens when we dodge a thunderbolt? A further Tim Prentice update

Weeks after Hurricane Irene came and went, I followed up with the brilliant sculptor down the road, Tim Prentice, on how he protected his kinetic outdoor sculptures. His preparations were small, it turns out. He moved a caged lobster made of sheet metal named Fred into his barn, even though Fred demanded to face the gale unassisted. Tim's prudence turned out to be unnecessary. The storm was a non-event in

Cornwall, if you ignore the power being out for two days. Some of our neighbors did get walloped, but nothing like the devastation to the north in Vermont.

Tim said, "We lost a few branches here and there, but that just means we have plenty of firewood for the winter." Fred has since moved out of the barn back to his usual post.

Were all the warnings about Irene justified? Luck steered the storm away from Carbon Based and our neighbors. We could have skipped all the preparations. We know that now, but we can't be sure what will happen the next time. Had the storm followed a slightly different track, we could have experienced Vermont-level destruction.

A risk that doesn't happen doesn't mean that the warnings were unnecessary. Of course, psychology being what it is, the next mega-storm may get short shrift because people will remember that Irene wasn't so bad in their neighborhood. They will think that way, even though Irene resulted in some $3 billion in losses.

This gap between direness of the warnings and the actual outcomes is endemic to discussions of long-term risks. How do we stay prepared for chronic dangers that will be around for decades-- like climate change? The classical environmental campaign, at least in caricature, involves scaring as many people as much as possible, to spur citizens' willingness to act. But in the long term, this is counterproductive. Most of us quickly get exhausted by overused alarms.

Climate change demands that we maintain our willingness to act for decades. We have to find ways of staying vigilant without triggering resistance in all of us, and without getting tired of the Cassandras.

In short, we have to learn to hover in the breeze like one of Tim's sculptures, and learn when to ask to be taken inside.

Previous page: Tim Prentice's open-air porch, with one of the worst-insulated windows ever. Photo by Brian Thomas

Thursday, September 29, 2011

My own failing dam

I run a story about a judge halting the Belo Monte dam project, but with total hypocrisy I ignore the ailing hunk of concrete and rock in my back yard. I'm talking about a literal dam. Shown here after today's soaking rain, this dam does an erratic job of blocking a stream that flows into the Housatonic River.

Another fact is obvious in this photo. The dam is falling apart. When we bought the house ten years ago, the structure was in pretty bad shape, a result of having skipped the rebar when pouring the concrete. A decade of wear has collapsed the spillway even more, and allowed more water to flow around the edges of the concrete walls.

Local beavers do figure-eights in the headwaters, waddle ashore, and plug holes where they can, tut-tutting at our lack of attention to the unconfined water. Their own dams are upstream, and the inhabited ones are in great shape.

A proper repair would require permissions from our local Inland Wetlands regulators, which would probably be doable. Then comes the expense. A minimal job would cost tens of thousands of dollars. The Architectural Digest version would cost several times that at least. I've done dozens of drawings of artful concrete structures, and then set them aside.

For now, we're letting it fail, and enjoying the sound of rushing water.

Sunday, October 30, 2011

Severed from the grid and the Internet

Nothing beats an intensified water cycle for disrupting your blog. Almost two feet of snow have buried Carbon Based. We are currently limping along with generator power and only a trickle of Internet. Posting will be light for a few days.

[Posted with iBlogger from my iPhone]

Thursday, November 3, 2011

Back from the void

Our power and internet just returned, after we went dark during the Halloween nor'easter.

Here we are, cooling our heels at a local school serving as a warming center, charging our phones and tablets. I haven't gotten the hang of blogging from an iPad, so I've been reading *Anna Karenin* instead. It feels appropriate, since Litchfield County has had a 19th century Russian feel to it, only with internal combustion engines and Facebook.

Our personal adaptation to climate disruption has been mixed. On the one hand, we shelled out a few years ago for a powerful generator that kept us going for two and a half days. Then the thrumming fell silent, plunging us into gloom. The lubricating oil had gotten used up, and the manual made refilling it seem daunting. After two dark, cold days, the local electrician's office called and explained how to do it myself. Back in business. I'll be ready next time.

And we were among the fortunate ones. We talked to one guy from Norfolk who had to get up early, chop enough wood for himself and his parents, and then rush to work. At the local school, elderly and disabled neighbors were warming up.

We're hearing much justified muttering about our utility cutting back on trimming trees and on maintenance in general. But still -- two big storms in the past two months strains the entire system.

Links from the blog Carbon Based

Monday, January 3, 2011

Natural disasters 'killed 295,000 in 2010'

AFP: Haiti earthquake and floods in Pakistan and China helped make 2010 an exceptional year for natural disasters, killing 295,000 and costing $130 billion, the world's top reinsurer said Monday. "The high number of weather-related natural catastrophes and record temperatures both globally and in different regions of the world provide further indications of advancing climate change," said Munich Re in a report.

The last time so many people died in natural disasters was in 1983, when 300,000 people died, mainly due to famine in Ethiopia, spokesman Gerd Henghuber told AFP. A total of 950 natural disasters were recorded last year, making 2010 the second worst year since 1980. The average number of events over the past 10 years was 785.

And in terms of economic cost, insured losses amounted to approximately $37 billion, putting 2010 among the six most loss-intensive years for the insurance industry since 1980. "2010 showed the major risks we have to cope with. There were a number of severe earthquakes. The hurricane season was also eventful," said Torsten Jeworrek, the firm's chief executive.

…The American continent suffered the most disasters -- 365 in total -- with 310 in Asia. A total of 120 natural disasters were recorded in Europe, 90 in Africa and 65 in Australia and Oceania. In 2009, considered a "benign" year due to the absence of major catastrophes and a less severe than usual hurricane season in the North Atlantic, there were 900 "destructive natural hazard events", costing some 60 billion dollars….

A composite image of the 2010 hurricane season

Wednesday, January 12, 2011

Interaction of knowledge, risk perceptions and action for climate events

WOMAN'S HOLY WAR.
Grand Charge on the Enemy's Works

[Baylor University](): New results from a Baylor University study show that different behaviors and strategies lead some families to cope better and emerge stronger after a weather-related event. Dr. Sara Alexander, an applied social anthropologist at Baylor who conducts much of her research in Central America, studied different households in several coastal communities in Belize. While climate change has been an emerging topic of interest to the world community, little scientific data exists on exactly how people respond to different climate-related "shocks" and events such as more intense hurricanes and prolonged drought.

Using a livelihood security approach, Alexander and her team identified vulnerable households in these communities and examined how they adapted and coped with major climate events and shocks such as droughts, hurricanes and floods. The Baylor researchers also developed tools to measure each household's long-term resilience, an area that has not been extensively researched, and identified specific behaviors and strategies that allowed some families to "weather the storm" better than others. The results indicate:

- …Perception about climate change and weather patterns played a key role in determining whether a household prepares adequately for a harsh weather event. For instance, 57 percent of households believed that storms today are more intense than they were five to 10 years ago, the household is more likely to prepare when weather forecasters predict threatening weather.
- Vulnerable and more secure households differ in coping strategies when dealing with weather-related events. Forty-nine percent of vulnerable households turn to their faith, 43 percent to their family, and 36 percent turned to their friends for emotional support. Only 19 percent turned to financially-based responses and only 8 percent made attempts to secure credit to gain resources to make repairs or rebuild. Households that have the highest levels of security are more likely to use their savings or sell their assets to engage in a financially based response by repairing and rebuilding, many times finding emotional support through this work.
- A critical ingredient for reducing vulnerability and enhancing resilience is empowerment of marginalized groups and the associated access to resources.
- Although the capacity of households to adapt to harsh weather is a function of perception of risk and access to resources, resilience of communities depends on the ability of people to think and act collectively….

An allegorical 1874 political cartoon print, which somewhat unusually shows temperance campaigners (alcohol prohibition advocates) as virtuous armored women warriors (riding sidesaddle), wielding axes Carrie-Nation-style to destroy barrels of Beer, Whisky, Gin, Rum, Brandy, Wine and Liquors, under the banners of "In the name of God and humanity" and "Temperance League". The foremost woman bears the [shield]() seen in the Seal of the United States (based on the U.S. flag), suggesting the patriotic motivations of temperance campaigners.

Sunday, January 16, 2011

In Ventura, a retreat in the face of a rising sea

[Tony Barboza in the Los Angeles Times](#):
At Surfers Point in Ventura, California is
beginning its retreat from the ocean.
Construction crews are removing a
crumbling bike path, ripping out a 120-
space parking lot and laying down sand
and cobblestones. By pushing the asphalt
65 feet inland, the project is expected to
give the wave-ravaged point 50 more
years of life.

The effort by the city of Ventura is the
most vivid example to date of what may
lie ahead in California as coastal
communities come to grips with rising sea
levels and worsening coastal erosion. As
the coastline creeps inland, scouring sand
from beaches or eating away at coastal bluffs, landowners will increasingly be forced to decide
whether to spend vast sums of money fortifying the shore or give up and step back.

State officials say the $4.5-million project in Ventura is the first of its kind in California and
could serve as a model for threatened sites along the coast. "Managed retreat, as it's called, is one
of the things that we're going to have in our quiver to deal with sea-level rise and increasing
storms," said Sam Schuchat, executive officer of the California Coastal Conservancy, which
helped fund the Surfers Point project.

…For years, the preferred solution to an eroding shoreline has been to build sea walls or dump
imported sand to serve as a buffer. About one-third of the Southern California coastline and
about 10% of the shore statewide have been fortified with sea walls and other hard structures.

Although artificial barriers may protect property in the short term, they often intensify the effect
of waves, leaving beaches stripped of sand until they narrow or disappear, permanently altering
surf patterns. As a result, beach-armoring projects are increasingly out of favor with
environmentalists and coastal regulators….

*A fisherman at the Ventura Pier, shot by [Randy](#), Wikimedia Commons via Flickr, under the [Creative Commons](#)
[Attribution 2.0 Generic](#) license*

Thursday, February 10, 2011

Extreme weather batters the insurance industry

A few snips from a long article by Ben Berkowitz in Reuters: …One of the biggest problems for insurers is that they have to insure increasingly valuable properties in risky areas that, by and large, are not being built with disaster risk in mind. That in and of itself is driving their risk up dramatically. When an insurer writes a policy for a property, it takes various factors into account, such as the property's location, its age, the propensity of the region it's in to be affected by weather events and the potential cost of replacing the property if it is damaged or destroyed.

Those criteria have largely stayed the same over the years, but what changed is the value of the properties to be insured and the volume of them. People around the world love beachfront houses and developers love selling them. In most places, no one stopped to think whether building the houses was a good idea, or whether there were appropriate building codes in place, or how many billions of dollars would be at stake if a major hurricane blew through.

"Even if the baseline of activity from a natural hazards point of view stays constant, the level of losses you're going to see will certainly be increasing commensurate to the increases in economic activity and national wealth," said Bill Keogh, the president of Eqecat, another major global risk modeler.

Most people in the business of predicting risk agree with Keogh that the changes in the environment matter less now than the changes in the "built environment" -- the size, value and type of buildings being put in high-risk areas like Florida's coastal zone and geologically unstable areas of California. Stringent building codes would overcome much of those risks, but such things either do not exist or are not strictly enforced in many parts of the world, and even in the United States they are a state-by-state patchwork. In many cases, it takes a disaster for them to be updated to reflect modern demands….

A house destroyed by Hurricane Ike in Shoreacres, Texas, 2008. Photo by FEMA

Thursday, February 17, 2011

Helping businesses adapt to climate change

Ryan Schuchard and Joyce Wong of the consulting firm BSR weigh in at Climate Biz: As climate change sets in, its impacts -- increasing severity of storms and weather disasters, receding snow and rivers, advancing deserts, and morc frcqucnt landslides and floods — will test companies' ability to effectively deliver high-quality products and services.

In response, BSR is launching a series of briefs to illustrate how these changes will affect each industry and what current adaptation practices look like, beginning an examination of the food, beverage, and agriculture sector.

Some effects of climate change will be familiar, such as crop failures and ensuing price shocks, but over the next several years, they will happen with more frequency and with even higher insurance costs. Beyond direct business impacts, companies will also need to understand how climate change will affect their most vulnerable stakeholders -- the poor, citizens of developing countries, and women -- who will face increasing risks due to drought, disease vectors, and the perils of migration.

The good news is that many resources on business adaptation to climate change are already available ... McKinsey & Company developed a cost curve for adaptation, for example, which highlights different adaptation options and shows that investment paybacks can be short. Also, companies do not need to choose between adapting to climate change and helping to mitigate it; the distinction between these two is rarely clear and we should do both together.

There are also tools that translate state-of-the-art climate monitoring, prediction, and imagery into practical information to help companies improve their relevant governance and decision-making processes. These tools include: the Climate Administration Knowledge Exchange (CAKE), Google Earth Engine, the International Research Institute for Climate and Society, the National Oceanic and Atmospheric Administration's Climate Prediction Center, and weADAPT. Companies can also take advantage of new market opportunities by providing solutions to enable effective adaptation…

In April 2009 -- Pat Troal, a Fargo, ND engineer marks a stake being used to measure the height of the river in the Edgewood neighborhood in Fargo. The city is measuring the height of the Red River to ensure that existing levees are high enough to protect the city. Photo by Patsy Lynch/FEMA

Saturday, March 19, 2011

Rise in disasters triggers review of risks

Ini Salgado in Business Report (South Africa): The catastrophe in Japan triggered by last week's earthquake, tsunami and nuclear scare underlined the need to deal with the rise in risks from natural disasters, short-term insurer Santam's head of strategy, Vannessa Otto-Mentz, said this week. "Before Japan it was the earthquake in Christchurch, the floods in Queensland, the Russian fires and drought, the Vaal and Orange

River floods," she said. "It punctuates the fact that risks are increasing, and something needs to be done."

German reinsurer Munich Re's latest disaster report counted 950 natural disasters worldwide last year, compared with an annual average of 785 over the past decade and 616 over three decades. Major global catastrophes killed 295,000 people last year and caused losses worth $130 billion (R918bn), nine-tenths of them due to severe weather-related events.

In Johannesburg this week, the UN Environment Programme (UNEP) Finance Initiative kicked off a consultation process around a set of principles for sustainable insurance to deal with environmental, social and governance risk, aimed at facilitating different approaches to underwriting, claims management and investment.

Because the industry was going to have to pay out greater claims, it would have to shift from its role as a risk carrier to a risk manager, said Deon Nel, the head of ecological research at the Council for Scientific and Industrial Research, who is conducting research for Santam on natural disasters in the Eden district. Nel's initial findings indicate that changes to the district's ecological buffering capacity, such as the removal of wetlands for agriculture, have at least as much impact on river zones as severe weather events.

Nel believes the insurance industry can deal with increased risk either by designating arbitrary boundaries beyond which it will not insure, or by adopting a model of shared risk that incentivises policyholders to become proactive risk managers. Otto-Mentz, a member of the working group that drafted the UNEP sustainable insurance principles, said South Africa could learn from and apply Japan's preparedness for earthquakes to its own risks, including those arising from floods, fires and dolomitic damage on the Witwatersrand....

Flooded level, West Driefontein mine. The dolomite overlying the gold reefs hosts much groundwater, giving the mines flooding problems. Shot by Babakathy, *who has released the image into the public domain*

Thursday, April 21, 2011

How our designs must adapt to a hotter world

Fastcodesign: ...Envisioning what the climate will be like 50, 75, or 150 years out is the first step in adaptation planning. We then need to ask: Which activities will be affected and how? What policies will help us cope with predicted change in climate? What opportunities will emerge? How will the design of our cities interface with our new climate? We need to alleviate predicted negative impacts through systems-thinking and build resilience into the system.

Adaptation is more than a good idea. It is a must --

especially when considering that many populations that are most affected by climate change tend to have fewer resources with which to cope with changes -- necessitating the design of adaptable communities as soon as possible. In other words, moving to higher ground is not as easy as it sounds.

…The better we plan now, the easier it will be for us to sustain and prosper in the future. This new way of thinking spans the design, policy, and even global market trends for businesses. While it may not be a national priority at the moment, many local governments are already feeling the impact of these changes and preparing themselves to adapt to continued change in the future through integration with the development of long-range plans. For example, the Coast Guard has moved lighthouses and the Massachusetts Water Resources Authority elevated a water treatment facility -- both projects stemming from predicted sea level rise and impact on infrastructure.

Adaptation should not be considered a plan in lieu of reducing emissions, but as a parallel effort. The good news is that many strategies exist to help communities and our designs adapt and reduce emissions, by focusing on "using less" and working within real world constraints on natural resources and energy security. For example, if our plumbing fixtures use less water, then we are not as greatly affected by drought and if we move toward distributed renewable energy, then we will be less threatened by energy distribution interruptions…

Shot of Andy Goldsworthy's Storm King Wall by Brian Thomas, under the Creative Commons *Attribution-Share Alike* 3.0 Unported

Wednesday, May 11, 2011

Finance, insurance, construction sectors announced joint private sector commitment to tackle disaster vulnerability

A report drawing on a panel at the disaster reduction conference underway in Geneva, from the United Nations International Strategy for International Disaster Reduction: Representatives of the finance, insurance and construction sector at the Global Platform yesterday sounded a call for action on the five essentials for business in disaster risk reduction, in line with the main finding of the United Nations Global Assessment Report on Disaster Risk Reduction that losses from disasters are rising faster than gains made through economic growth.

The Global Platform – held in Geneva every two years -- is the world's main forum for disaster risk reduction, attracting nearly 3,000 participants this year and includes a bigger private sector contingent than the previous Platform. The Platform opened yesterday and ends on Friday, 13 May.

In an announcement to the press, WillisRe, Credit Suisse, Titan America, Cisco Internet Business Solutions Group and MunichRe called on fellow members of the private sector to: build partnerships to analyze the root causes of non-resilient activity; leverage private sector expertise in construction, communications, financing, transport and contingency planning; spread knowledge about risk, prediction, forecasting and early warning; assist governments to conduct risk assessments; and help develop standards and procedures for enhancing resilience.

"We commit voluntarily and to the best of our abilities to create awareness within and outside our organizations to identify vulnerability and their root causes in our areas of activity and influence," said Peter Gruetter, CISCO Internet Business Solutions Group, adding that the group recognized the leading role of the International Strategy for Disaster Reduction – established by the United Nations in 2000 to consider actions for reducing risk – and the Hyogo Framework for Action, which is the world's only internationally-recognized blueprint for disaster risk reduction.

According to a report put together by the United Nations secretariat for the International Strategy for Disaster Reduction, which was launched Tuesday at the 2011 Global Platform for Disaster Risk Reduction, disaster-related losses are increasing across all regions, threatening the economies of low- and middle-income countries as well as outpacing wealth gains across many of the world's more affluent nations…

The caption from Wikimedia Commons reads: "Looking North from Ursuline Academy, showing wrecked Negro High School Building, Galveston, Texas." A hurricane did the wrecking

Sunday, May 15, 2011

Flooding may end the stalling for FEMA maps

Mike Fitzgerald in the Belleville News-Democrat (Illinois) covers an issue that's going to keep arising – the unsoundness of building in flood-vulnerable areas. It's going to be a farrago of corruption, wishful thinking and malfeasance as homeowners and politicians attempt to argue the risks away: Metro-east leaders have for years succeeded in playing for time when it came to new flood-hazard maps for Madison, St. Clair and Monroe counties. With the help of federal lawmakers, they persuaded the Federal Emergency Management Agency again and again to push back deadlines for implementing the long-dreaded maps -- dreaded because they would sharply increase the number of property owners forced to buy flood insurance, as well as the cost of premiums to obtain it.

But now, with record floods overwhelming towns along the lower Mississippi River, it looks unlikely that FEMA -- and leaders of Congress -- will want to budge beyond the February deadline for the new flood maps. Such is the gloomy assessment of Les Sterman, the chief engineer for the Southwestern Illinois Flood Prevention District Council.

Faced with almost $18 billion in debt arising from Hurricane Katrina and its aftermath in 2005, FEMA needs to rake in new revenue by extending the zones where property owners must buy federal flood insurance, according to Sterman.

"There are members of Congress who want it to stand on its own," said Sterman, who is overseeing the $130 million effort to rebuild the metro-east's levee system. "It can't do that without collecting more insurance premiums from other folks, particularly other folks like us who'll probably never flood."

U.S. Sen. Dick Durbin, D-Springfield, disagreed with Sterman's assessment. In March, Durbin announced that FEMA had finally agreed to a request by Durbin and other senators to recognize the existence of flood-protection levees in the metro-east and elsewhere nationwide.

…For Durbin, the issue of flood insurance represents a swinging political tight rope. On one hand, Durbin realizes the new FEMA flood maps are extremely unpopular. Government and business leaders predict they could trigger economic calamity for 150,000 American Bottoms property owners in the region, many of whom live in some of the state's poorest neighborhoods.

Sharply higher insurance premiums would force many to leave the area, while stifling badly needed economic development. But Durbin has also grown worried about the increased frequency of 100- and 500-year flood events brought on by climate change, according to Mulka. Durbin "wants to make sure that areas are prepared, whether that means protecting yourself with flood insurance or upgrading the levees," Mulka said....

An aerial view of Alton in Madison County, Illinois, taken June 20, 2008, shows flooding of the Mississippi River. US Air Force photo

Sunday, May 22, 2011

Americans take a gamble with the Mississippi floods

Suzanne Goldenberg in the Guardian (UK) describes a deeply ingrained US tendency to ignore risk. Since preventing development in dangerous areas rarely has much standing on the nation's agenda, we keep making the same mistake for the sake of short-term gain: …Seen one way, the floods are an act of nature, beyond human control – and America got off relatively lightly. Despite extensive property damage, with predictions that more than a million acres of land would go underwater, only four deaths due to flooding have been recorded so far, in Arkansas and Mississippi. Industries, population centres and shipping in the Mississippi have been protected. Unlike in Hurricane Katrina, the Army Corps of Engineers, which is in charge of flood control, has had no levee failures.

Looked at another way, the flooding was entirely predictable. Damage to homes and fields in the Mississippi's way should have been avoidable. For all the effort over the years put into controlling the Mississippi, for many individuals – and even entire towns – there is only the illusion of safety.

…There is complacency and resignation. Flood experts argue that America's flood protection standards are lax compared with those in other countries. Authorities have hesitated to relocate people to safer ground, or to enforce laws that compel local authorities to provide flood protection and require homeowners to get flood insurance.

"We have been very good at letting people continue to live in harm's way," said George Galloway, who was commander of the Army Corps at Vicksburg in the 1970s. "But how much longer can we continue to do that since we know with climate change we are going to have more floods than in the past?" In the 1990s, Galloway led a White House study into improving flood protection. It concluded that most people living in flood areas – up to 7 million across the country – did not fully understand the risks they faced.

…It is becoming evident that the Army Corps of Engineers and other forecasters have underestimated the frequency of severe flooding along the Mississippi. "We had a 500-year flood in 1993, a 70-year flood in 2001, and a 200-year flood in 2008. What blows my mind is that I just published this paper in 2008 and every year since then we have had another 10-year flood," said Robert Criss, a hydrologist at Washington University in St Louis. "The observed frequency of flooding is completely incompatible with the Army Corps estimates."…

The Great Mississippi Flood of 1927, Greenville, Mississippi. The river stage was at 46.8 feet. From: "The Floods of 1927 in the Mississippi Basin", Frankenfeld, H.C., 1927 Monthly Weather Review Supplement No. 29.

Wednesday, May 25, 2011

The Colorado River's (nonexistent) emergency plan

[Rob Davis in the Voice of San Diego points out a hole in the governance of the Colorado River's water](): The Colorado River, the lifeblood water supply of San Diego and the Southwest, made history late last year. And it wasn't good. Lake Mead, the reservoir holding the Colorado River back behind the Hoover Dam outside Las Vegas, hit its lowest point since being filled in the 1930s. Had it dropped just a few more feet, federal officials would've declared the first shortage in the river's modern history. Arizona and Las Vegas would've gotten less water.

Fortunately, the lake's decline stopped. Officials began a massive transfer of water from Lake Powell, the river's other major reservoir, on the Utah-Arizona border. Along with a wet winter in the Colorado Rockies, that staved off what was once unthinkable. The Colorado River wasn't supposed to come up short.

Though the Southwest got a reprieve, it was a troubling sign for a river relied on by 27 million people in places like San Diego, Los Angeles, Phoenix, Las Vegas and across the border in Mexico. The Colorado River today faces two serious threats, either of which would mean less water for the arid Southwest.

The first: The river's annual supply was divided in the 1920s during a historically wet period. The Southwest created its expectations of the Colorado's annual yield during one of the wettest centuries on record. If yields return to their historic average, there won't be enough water to go around.

The second: Climate change is projected to cut the Colorado's yield between 10 percent and 30 percent by 2050. Even a small reduction would have a major impact. In the lower Colorado basin, California, Arizona, Nevada and Mexico take nearly every last drop to which they're entitled. Scientists at the Scripps Institution of Oceanography say a 10 percent drop could cause shortages in six of every 10 years.

Underlying those dire projections is a major uncertainty: If the Colorado consistently comes up short, no one knows who will cut consumption to keep Lake Mead from running dry. A short-term plan is in place from the time the reservoir hits 1,075 feet above sea level until 1,025 feet. Arizona and Las Vegas take that hit. (At 895 feet, the reservoir wouldn't be able to distribute water and the Hoover Dam wouldn't produce power. The lake is currently at 1,096 and climbing.)…

Thursday, June 2, 2011

The role of insurers in climate change adaptation

[Claudia Woo at CSR Asia](#): …Not all disasters are related to climate change, but the increasing occurrence of natural disasters due to climate change has indeed placed a huge financial burden on governments worldwide to provide immediate aid to victims and to fund rebuilding efforts in the aftermath. With less capacity and economic resources to alleviate or adapt to climate change, developing nations are particularly vulnerable to natural catastrophes. It is estimated that the cost of climate change to emerging economies is between one and 12 per cent of annual GDP. This figure could surge to 19 per cent under severe climate conditions, according to the Economics of Climate Adaptation (ECA) Working Group.

Regardless of the efforts to control greenhouse gases emissions, it is increasingly apparent that adaptation to climate change is necessary, including seeking ways to finance economic losses after a climate disaster occurs. Climate-related micro-insurance, which usually charges a low premium for the poor, is a cost-effective means of promoting climate change adaptation measures. It indemnifies low-income groups against specific losses due to climate hazards.

For instance, an insurance scheme has been implemented in Ethiopia under the Horn of Africa Risk Transfer for Adaptation (HARITA) project to compensate local farmers growing teff and are affected by drought. The scheme is unique because instead of using money, local farmers can pay insurance premiums by labouring on projects that will reduce climate change effects in their communities, such as tree planting….

Other examples of climate-related micro-insurance schemes can be found in Kenya (index-based weather insurance), Caribbean nations (parametric insurance against hurricanes), and Mexico (catastrophe bonds). Such insurance systems can be adopted or adapted to other nations by taking into consideration location-specific climate conditions.

…Nevertheless, penetration of catastrophe-related insurance is still extremely low in developing countries in Asia, For example, it is still under five per cent in India, the Philippines and China. Many people tend to ignore the risk of natural perils due to the low frequency rate, but such attitudes may change over time given the increased incidence of climate disasters globally….

Fire insurance marks on display at Bedford Museum. Shot by [Simon Speed](), Wikimedia Commons, public domain

Tuesday, June 14, 2011

Can the insurance industry survive climate change?

[Francesca Rheannon in Reuters]():
…[M]ost victims of extreme weather disasters are more likely to call on their insurance companies instead of the Lord for relief. So far, homeowners in western Massachusetts have filed insurance claims totaling a record $90 million with the final number expected to rise. The claims from wildfires, floods and tornadoes are likely to set records in other parts of the country, as well. The Joplin, MO tornado alone will cost insurers an estimated $3 billion, with the total for the tornadoes that ravaged the Midwest expected to top $7 billion.

…But how will the industry respond? The most immediate response we are already seeing are soaring premiums to homeowners and businesses. One 2009 study predicted a doubling of insurance rates due to climate change - and that was before severe weather events doubled in 2010 from 2009 totals.

Other companies are pulling out of climate-change challenged markets altogether, as the New York Times reported. "Allstate, the largest publicly traded insurer in the United States, has reduced its exposure in hurricane-prone areas and stopped writing new homeowners policies in California altogether, because of earthquake risk. It has also trimmed back its risk from inland storms." The report quoted one Allstate executive as saying, "We're running our business as if this change...is permanent...and we need to recover those costs."

Another way of recovering costs is by limiting coverage through increased deductibles, reduced limits and novel exclusions (after Katrina, some homeowners were denied coverage when their house flooded because it was claimed the damage came from the hurricane-force winds).

Insurance may well become unaffordable for millions in the U.S. and elsewhere in the developed world as premiums skyrocket. But the problem in the developing world is that many of the world's most vulnerable victims of climate change are not insured at all.…

US Navy photo of Hurricane Katrina damage to Gulfport, Mississippi, in 2005

Tuesday, June 28, 2011

Water use in China and the Middle East is an environmental Ponzi scheme

Damian Carrington in the Guardian (UK): Find water and you find life. This simple maxim guides scientists searching distant planets for aliens. But if the astrobiologists were to reverse their telescopes and look at our own globe, they would find a conundrum: billions of people living in places with little or no water.

That unsustainable paradox is now unravelling before our eyes in the Middle East and north Africa. The 16 most water-stressed states on Earth are all in that troubled region, with Bahrain at the top of the ranking from risk analysts Maplecroft. Libya, Yemen, Egypt, Tunisia and Syria follow not far behind.

All are built on an environmental Ponzi scheme, using more water than they receive: 700 times more in Libya's case. The unrest of the Arab spring of course has many causes, but arguably the most fundamental is the crumbling of a social contract that offered cheap water – and hence food – in return for subservience to dictators.

The region's population is rocketing – there are 10,000 new mouths to feed each day – just as grain production plummets. The deep, ancient aquifers that enabled crops to green the deserts are almost exhausted, and the oil that fires the desalination plants to make up the loss is dwindling too.

It's a perfect storm of water, food and energy crises and has arrived two decades sooner than even the most sober analysts expected. And while the Middle East is the first region to feel the wrath of that storm, across the world warning signs are flashing – from the sinking of Mexico City as its aquifers are sucked dry to the docking of freshwater tankers in Barcelona…

A ship in what used to be the Aral Sea, long since dried up. Shot by Staecker, who has released it into the public domain

Saturday, July 9, 2011

Billion-dollar disasters on the rise

Kimberley Schupp in WAFF News: Tornadoes, earthquakes and hurricanes are devastating natural disasters that could affect anyone. According to Insurance Center Associates, these natural disasters are becoming more frequent and more costly. Prior to 1987, the U.S. never experienced a natural disaster with insured losses greater than $1 billion. Since that time, there have been eight.

A new report from Kiplinger reveals the 10 states most at risk of a natural disaster. The rankings are based on estimates of each state's insured property losses across the past decade. Kiplinger provides financial services and publishes magazines and newsletters. The group named Louisiana as the No. 1 state for natural disasters.

Experts say climate changes and global warming are to blame for the extreme weather that has pummeled the U.S. in 2011. "Any single weather event is driven by a number of factors, from local conditions to global climate patterns and trends. Climate change is one of these. It is very likely that large-scale changes in climate, such as increased moisture in the atmosphere and warming temperatures, have influenced, and will continue to influence, many different types of extreme events, such as heavy rainfall, flooding, heat waves and droughts," Thomas Karl, director of the National Climatic Data Center said.

According to the Pew Center on Global Climate Change, in the first six months of 2011 the U.S. experienced record-breaking floods and snowstorms, prolonged drought, and massive wildfires....

House destroyed by a 2011 tornado in Brimfield, Massachusetts, shot by Mass DEP, Wikimedia Commons via Flickr, under the Creative Commons Attribution 2.0 Generic license

Tuesday, July 12, 2011

Resilience of US metro areas measured by online index

University at Buffalo (SUNY) News Center: Which U.S. metro region is most likely to come out of the next recession, natural disaster or other regional "shock" relatively unscathed? Rochester, Minn. A little more battered might be College Station-Bryan, Texas. These two regions are ranked first and last, respectively, by a new online tool measuring more than 360 U.S. metros for their "regional resilience," or capacity to weather acute and chronic stresses ranging from gradual economic decline to rapid population gains to earthquakes and floods.

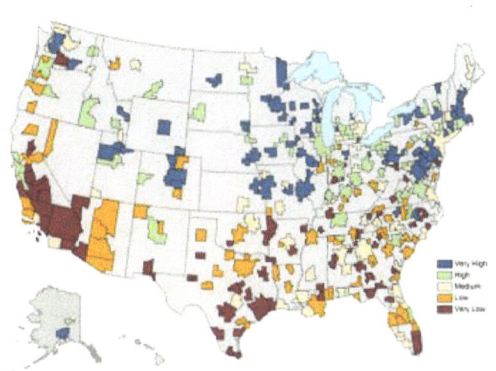

The Resilience Capacity Index (RCI), developed by Kathryn A. Foster, director of the Regional Institute, a research and public policy center of the University at Buffalo, produces a single statistic for each region based on its performance across 12 economic, socio-demographic and community connectivity indicators, ranging from income equality and business environment to voter participation and the population of health-insured.

As a gauge for how well a region is positioned to adapt to stress, the index can help regional leaders identify strengths and weaknesses and target related policy changes toward building their resilience capacity. "Conceiving of regions as capable of adapting and transforming in response to challenges allows researchers and practitioners to understand the conditions and interventions that may make one place more or less resilient and why," said Foster, also a professor of urban and regional planning with UB's School of Architecture and Planning.

Foster developed the index as part of Building Resilient Regions, a national network of experts on metropolitan regions funded by the John D. and Catherine T. MacArthur Foundation and administered by the University of California, Berkeley. Online at http://brr.berkeley.edu, the index features maps revealing geographic patterns in resilience capacity, detailed data profiles for each metro and a "compare metros" tool.

Overall, Northeastern and Midwestern regions tend to be more resilient than those in the South or West, largely because these regions earn high scores for affordability, the size of their health-insured population, rates of homeownership and metropolitan stability, as measured by recent population change....

UB researchers examined more than 360 U.S. metro areas to determine which would be most likely to come out of the next recession, natural disaster or other regional "shock" relatively unscathed. Image from the UB website

Monday, August 1, 2011

Insurance industry grapples with impact of climate change

Janet Whitman in CNBC writes about the pressure on insurance company profits, in a story that illustrates why the insurance business is so unforgiving. A company can write policies that look profitable year after year, only to have all those profits and more vanish in a big loss year like this one. Or, as an investment banker once told me, insurance companies are really hedge funds with an unfortunate addiction to writing insurance: As a surge in catastrophic weather events leads to billions of dollars in claims, climate change may pose the insurance industry's biggest problem — and profit potential. An unusual onslaught of floods in Mississippi, tornadoes in the Midwest, drought and wildfires in Texas and earthquakes abroad has wiped out hope of much profit for many insurance and reinsurance companies this year.

Natural disasters, including Japan's earthquake and tsunami, have left the industry on the hook for $60 billion in the first six months of this year alone, according to data from reinsurance giant Munich Re. That's nearly five times the first-half average since 2001, and the oftentimes costly hurricane season is just getting underway.

Although the damage has been worse than expected, the industry has plenty of capital set aside to weather the losses — barring the occurrence of more off-the-charts disasters, which insurers and reinsurers say may or may not be linked to global warming.

"Many companies that write large catastrophe risks, primarily reinsurance companies, are getting paid money in many years when nothing happens," says Matthew Rohrmann, an insurance industry analyst with Keefe, Bruyette & Woods. "There really haven't been that many devastating hurricanes over the past few years, for example, and reinsurers and insurers are still getting paid premiums. They may lose a lot of money in one year, but their pricing is based on many years."...

Turner's "Shipwreck of the Minotaur", oil on canvas, Calouste Gulbenkian Museum, Lisbon

Monday, October 3, 2011

Flood-risk review creates waves in Australia

[Clancy Yeates in the Sydney Morning Herald](): Insurance companies face an overhaul in how they handle flood risk, after an independent review considered sweeping changes designed to lift insurance cover against natural disasters. The Assistant Treasurer, Bill Shorten, was due to receive a final report on Friday from the National Disaster Insurance Review, chaired by the former banking regulator John Trowbridge.

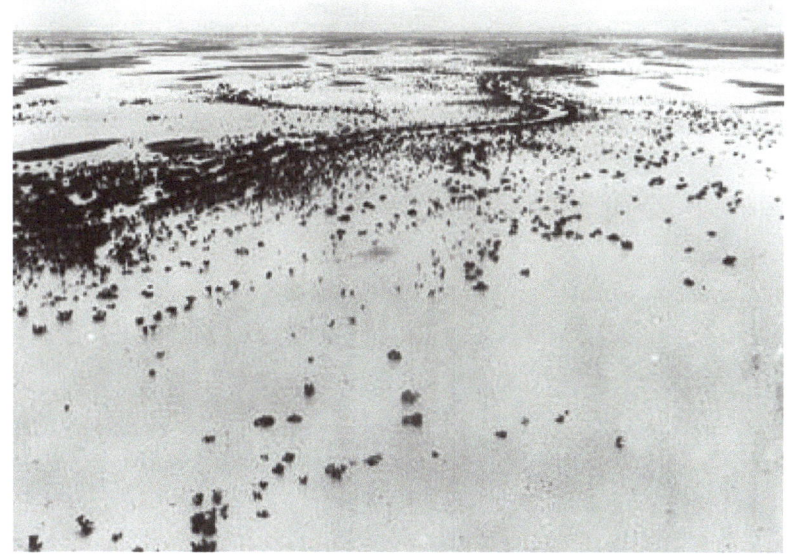

Industry sources said the review appeared to be leaning towards a proposal to create a new pool of funds that would be used to subsidise flood insurance in high-risk areas. It is unclear how such a pool would be funded, but some sources were concerned it could be paid for through a levy on insurance policies. The review has also considered other options to lift flood cover, which was included in about half of all home insurance policies last year. In a discussion paper, Mr Trowbridge raised the option of making flood cover an automatic part of policies, as storm cover is today.

Insurance companies are keenly awaiting the recommendations, but these are unlikely to be made public until Mr Shorten unveils the government's response later this year. As the review drew to a close in recent weeks, firms including the country's biggest general insurer, Suncorp, have railed against the prospect of further government intervention in the push to lift flood cover.

Insurers say creating an additional insurance pool for flood cover would push up costs for all consumers and distort the market. Because government revenue is under pressure from Canberra's pledge to deliver a budget surplus in 2012-13, industry sources said an insurance pool would probably be funded by a new levy.

Critics say this would mean policyholders in low-risk areas would be subsidising homes at high risk of flooding, eroding the incentive for home owners to protect themselves against flood risk....

Thomson River in flood at Jundah, Queensland, 1950.

Wednesday, September 14, 2011

Building codes may underestimate risks due to multiple hazards

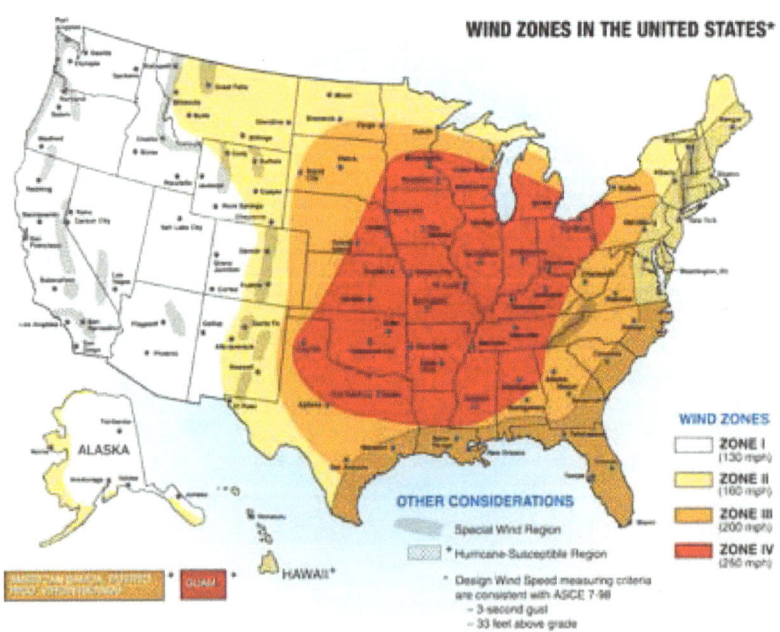

The National Institute of Standards and Technology (US Department of Commerce): As large parts of the nation recover from nature's one-two punch— an earthquake followed by Hurricane Irene—building researchers from the National Institute of Standards and Technology (NIST) warn that a double whammy of seismic and wind hazards can increase the risk of structural damage to as much as twice the level implied in building codes.

This is because current codes consider natural hazards individually, explains NIST's Dat Duthinh, a research structural engineer. So, if earthquakes rank as the top threat in a particular area, local codes require buildings to withstand a specified seismic load. In contrast, if hurricanes or tornadoes are the chief hazard, homes and buildings must be designed to resist loads up to an established maximum wind speed.

In a timely article published in the Journal of Structural Engineering, Duthinh, NIST Fellow Emil Simiu and Chiara Crosti (now at the University of Rome) challenge this compartmentalized approach. They show that in areas prone to both seismic and wind hazards, such as South Carolina, the risk that design limits will be exceeded can be as much as twice the risk in regions where only one hazard occurs, even accounting for the fact that these multiple hazards almost never occur simultaneously. As a consequence, buildings designed to meet code requirements in these double-jeopardy locations "do not necessarily achieve the level of safety implied," the researchers write.

Simiu explains by analogy: a motorcycle racer who takes on a second job as a high-wire performer. "By adding this new occupation, the racer increases his risk of injury, even though the timing and nature of the injuries sustained in a motorcycle accident or in a high-wire mishap may differ," he says. "Understandably, an insurer would raise the premium on a personal injury policy to account for the higher level of risk."...

Wind zone map shows how the frequency and strength of extreme windstorms vary across the United States. Wind speeds in Zone IV (red), where the risk of extreme windstorms is greatest, can be as high as 250 miles per hour. Credit: Federal Emergency Management Agency

Thursday, October 20, 2011

New report ranks deficient US bridges by metro areas

Transportation for America: A new look at structurally deficient bridges in metropolitan areas finds that just a quarter of U.S. bridges, located in our largest metropolitan areas, carry 75 percent of all traffic crossing a deficient bridge each day.

On the heels of the sudden closure of a major commuting bridge in Louisville, KY, a new report shows that more than 18,000 of the nation's busiest bridges, clustered in the nation's metro areas, are rated as "structurally deficient," according to this new report from Transportation for America.

In Los Angeles, for example, an average 396 drivers cross a deficient bridge every second, the study found. *The Fix We're In For: The State of Our Nation's Busiest Bridges*, ranks 102 metro areas in three population categories based on the percentage of deficient bridges.

The report found that Pittsburgh, PA had the highest percentage of deficient bridges (30.4 percent) for a metro area with a population of over 2 million (and overall). Oklahoma City, OK (19.8 percent) topped the chart for metro areas between 1-2 million, as did Tulsa, OK (27.5 percent) for metro areas between 500,000-1 million.

At the other end of the spectrum, the metro areas that had the smallest percentage of deficient bridges are: Orlando, FL (0.60 percent) for the largest metro areas; Las Vegas (0.20 percent) for mid-sized metro areas; and Fort Myers, FL (0.30 percent) for smaller metro areas.

"There are more deficient bridges in our metropolitan areas than there are McDonald's restaurants in the entire country," said James Corless, director of Transportation for America, 18,239 versus roughly 14,000 McDonald's. "These metropolitan-area bridges are most costly and difficult to fix, but they also are the most urgent, because they carry such a large share of the nation's people and goods."...

Saturday, October 22, 2011

Florida needs huge investment in water infrastructure

TCPalm (Florida): Florida Atlantic University science and engineering researchers within the Charles E. Schmidt College of Science and College of Engineering and Computer Science today released a study indicating that climate change will cause significant impacts on Southeast Florida's water infrastructure, attributable to sea level rise and growing variation in seasonal rainfall patterns with more intense periods of drought alternating with increased torrential rainfall events.

The research report, "Southeast Florida's Resilient Water Resources," and the case study titled "Improving the Resilience of a Municipal Water Utility" exemplified that — as a consequence of climate change impacts — Southeast Florida water utilities will face a number of challenges, including inundation of low-lying coastal areas; saltwater contamination of well fields; malfunction of septic tanks and drainage systems; reduction in soil capacity to store rainfall; and reduced efficiency of stormwater drainage canals and flood gates, among others. Strategies to manage these challenges would require substantial economic investments in the order of $500 million to $1 billion over the next 70 to 100 years. To support these improvements, household utility bills could increase by as much as $100 per month.

"Significant challenges to the water systems in Southeast Florida due to climate change are expected to begin within the next two decades. Water managers will have to contend with increasing saltwater intrusion and more intense drought. Furthermore, risk of flooding will increase as a result of more intense rain storms coupled with sea level rise that will cause reduced capacity of flood control systems," said Barry Heimlich, research affiliate with the FAU's Florida Center for Environmental Studies, who led the study. "Early notice of this study's findings helped raise awareness of these issues and encouraged regional water managers to incorporate climate change in water resource planning and begin development of flexible adaptation strategies to be implemented over the coming decades."...

Monday, November 7, 2011

How should society pay for ecosystems services?

University of Minnesota News: Over the past 50 years, 60 percent of all ecosystem services have declined as a direct result of the conversion of land to the production of foods, fuels and fibers. This should come as no surprise, say seven of the world's leading environmental scientists, who met to collectively study the pitfalls of utilizing markets to induce people to take account of the environmental costs of their behavior and solutions. We are getting what we pay for.

"The best things in life are free, including nature", says author Stephen Polasky, professor of applied economics and ecology, evolution and behavior. "But without a price for nature's services we don't maintain the environment in ways necessary to sustain these valuable services." Polasky is also a resident fellow in the university's Institute on the Environment (IonE). Their report, "Paying for Ecosystem Services: Promise and Peril," is published in the Nov. 4 issue of the journal Science.

Society pays for the products of agriculture, aquaculture and forestry, and has developed well-functioning markets for these products, these experts say. However, author David Tilman, Regents Professor in the College of Biological Sciences and IonE fellow, notes "We also need market mechanisms that reward farmers for the quality of the water that leaves their lands, and for other important ecosystem services." These services include watershed protection, habitat provision, pest and disease regulation, climate regulation and storm buffering.

The problem is that many ecosystem services are public goods. Some lie outside the control of any one government, and the science for others is still only poorly understood. There is no one-size payment mechanism that fits all cases. And bad payment mechanisms can be worse than no payment mechanisms at all, the study's authors warn, pointing to the lessons learned from four decades of agricultural subsidies. Subsidies encouraged the overuse of fertilizers and pesticides, two of the main reasons for the growing number of dead zones in the world's oceans...

Saturday, November 19, 2011

Study finds Great Plains river basins threatened by pumping of aquifers

Oregon State University News: Suitable habitat for native fishes in many Great Plains streams has been significantly reduced by the pumping of groundwater from the High Plains aquifer – and scientists analyzing the water loss say ecological futures for these fishes are "bleak."

...Unlike alluvial aquifers, which can be replenished seasonally with rain and snow, these regional aquifers were filled by melting glaciers during the last Ice Age, the researchers say. When that water is gone, it won't come back – at least, until another Ice Age comes along.

"It is a finite resource that is not being recharged," said Jeffrey Falke, a post-doctoral researcher at Oregon State University and lead author on the study. "That water has been there for thousands of years, and it is rapidly being depleted. Already, streams that used to run year-round are becoming seasonal, and refuge habitats for native fishes are drying up and becoming increasingly fragmented."

Falke and his colleagues, all scientists from Colorado State University where he earned his Ph.D., spent three years studying the Arikaree River in eastern Colorado. They conducted monthly low-altitude flights over the river to map refuge pool habitats and connectivity, and compared it to historical data.

They conclude that during the next 35 years – under the most optimistic of circumstances – only 57 percent of the current refuge pools would remain – and almost all of those would be isolated in a single mile-long stretch of the Arikaree River. Water levels today already are significantly lower than they were 40 and 50 years ago. Though their study focused on the Arikaree, other dryland streams in the western Great Plains – comprised of eastern Colorado, western Nebraska and western Kansas – face the same fate, the researchers say....

The Gunnison River at twilight, from Sunset View, Black Canyon of the Gunnison National Park, Colorado. Shot by Phil Armitage

Sunday, December 18, 2011

Virginia residents oppose preparations for climate-related sea-level rise

Why not let the denialist towns in Virginia serve as control groups for the question, what happens when vulnerable coastal areas make no attempt to adapt? Darryl Fears in the Washington Post: Over his long career as a public planner, Lewis L. Lawrence grew accustomed to the bland formalities of planning commission meetings in Virginia's Middle Peninsula, where forgetting to cover one's mouth while yawning through a lecture was about as rude as people got.

But lately, the meetings have gotten far more exciting — in a bad way, said Lawrence, acting executive director of the Middle Peninsula Planning District Commission. A well-organized and vocal group of residents has taken a keen interest in municipal preparations for sea-level rise caused by climate change, often shouting their opposition, sometimes while planners and politicians are talking.

The residents' opposition has focused on a central point: They don't think climate change is accelerated by human activity, as most climate scientists conclude. When planners proposed to rezone land for use as a dike against rising water, these residents, or "new activists," as Lawrence calls them, saw a trick to take their property.

"Environmentalists have always had an agenda to put nature above man," said Donna Holt, leader of the Virginia Campaign for Liberty, a tea party affiliate with 7,000 members. "If they can find an end to their means, they don't care how it happens. If they can do it under the guise of global warming and climate change, they will do it."

Outside of greater New Orleans, Hampton Roads is at the biggest risk from sea-level rise of any area its size in the United States, according to the National Oceanic and Atmospheric Administration. The water has risen so much that Naval Station Norfolk is replacing 14 piers at $60 million each to keep ship-repair facilities high and dry....

From left uppermost, going clockwise: Hampton, Norfolk, Portsmouth, Suffolk, and Newport News, Virginia, USA. - July 1996 One can see the major roadways throughout Hampton Roads

www.ingramcontent.com/pod-product-compliance
Lightning Source LLC
Chambersburg PA
CBHW051045180526
45172CB00002B/524